£11.95

D0585045

WITHDRAWN

FREE-RANGE POULTRY

FREE-RANGE POULTRY

Katie Thear

FARMING PRESS

First published 1990

Copyright © 1990 Katie Thear

British Library Cataloguing in Publication Data

Thear, Katie *1939–*
  Free-range poultry.
  1. Poultry. Care & management
  I. Title
  636.5083

  ISBN 0-85236-190-4

Published by Farming Press Books
4 Friars Courtyard, 30–32 Princes Street
Ipswich IP1 1RJ, United Kingdom

Distributed in North America
by Diamond Farm Enterprises,
Box 537, Alexandria Bay, NY 13607, USA

Cover design by Andrew Thistlethwaite
Phototypeset by Galleon Photosetting, Ipswich
Printed and bound in Great Britain by
Biddles Ltd, Guildford and King's Lynn

CONTENTS

v

PREFACE

This book is a practical and comprehensive guide to the free-range management of poultry, on any scale. It is up to date in its coverage, with an emphasis not only on how traditional practices have been adapted for modern conditions, but also on the role of scientific and veterinary developments in the avoidance and control of problems. It is not a book which looks nostalgically at the past, but gives a realistic appraisal of one of the most important innovations in the poultry industry in recent decades.

Public interest in more natural and unadulterated foodstuffs has combined with a greater concern for humane livestock management systems to produce a consumer demand for free-range eggs and for non-intensively reared table birds. Alternatives to the battery and broiler systems represent only a small proportion of production, but it is a growing area and one which is arguably indicating the path of future developments.

I hope that the book will prove useful to all potential and existing free-range producers, large and small, to agricultural students, to those catering for the needs of producers and, indeed, to anyone with an interest in poultry.

KATIE THEAR
Newport, 1990

ACKNOWLEDGEMENTS

I am grateful to the following individuals and organisations for their invaluable contributions and advice:

Dr Linda Keeling, the West of Scotland Agricultural College.
Mr R. G. Wells, National Institute of Poultry at Harper Adams College of Agriculture.
Mr George Hall, Agricultural Training Board.
Mr John Portsmouth, Peter Hand Animal Health.
Mr D. B. Brown, Salsbury Laboratories.
Mr L. A. Wright, Intervet Limited.
Mr Mick Dennett of Bibby Agriculture.
ADAS Livestock Services.
The National Agricultural Centre's Poultry Unit at Stoneleigh.
Poultry World magazine.
Farmers' Weekly magazine.

Finally, some of the material in this book has appeared in various articles in *Home Farm* magazine. I am grateful to the publishers for allowing it to be used again.

FREE-RANGE POULTRY

INTRODUCTION

'Preference is given to the genuine farm egg, produced by healthy stock enjoying free-range and living under natural conditions.'
                                                    – Herbert Howes, 1939

The quotation above arguably sums up the general view of free-range poultry-keeping. The concept of healthy, contented birds, ranging at will over natural vegetation and laying fresh farm eggs, is a cosy one. There is also a beguiling implication that consumers will give preference to free-range eggs over those produced in more intensive systems. Overall, it is an appealing picture – but how true is it?

It would be a mistake to assume that any system of husbandry is 'natural'; there is really no such thing. Where poultry are kept for production purposes, in field, orchard or back garden, there is inevitably a degree of interference by man. What it is true to say is that some systems are less intensive than others. A free-range system, where birds are allowed reasonably unrestricted access to grass paddocks, is obviously less intensive than a battery system, where birds are caged in a totally artificial environment, but it may be far more labour-intensive in terms of the man-hours required to run it.

Healthy birds are not necessarily those which have access to free-ranging conditions. Indeed, there are many diseases which are more prevalent under these conditions than in the batteries. The resurgence of commercial free-range production in recent years has also resurrected traditional problems such as Marek's disease, avian tuberculosis and internal parasites, problems which had been virtually eliminated in the batteries. This is not to suggest that batteries are free of health problems, merely that they have different ones. Salmonella infection is a condition which can affect both battery and free-range birds, and it has major implications for the health of the consumer.

1

When it comes to consumer choice, there is little doubt that most people opt for the cost factor and buy the cheapest eggs and table birds, and these are battery and broiler-produced. What is also true – and this is the salient factor for free-range producers – is that a growing minority of consumers are prepared to pay a premium price for free-range eggs and non-intensively raised table birds.

Where chickens are kept primarily for interest, it is often because there is a wish to have one's own fresh, quality produce rather than to buy older, intensively produced eggs or birds.

Adopting a critical standpoint in relation to free-range production does not imply a negative attitude to it. On the contrary, it makes good sense to be a 'devil's advocate', for only then is it possible to take an overview and remain reasonably objective. To start a poultry enterprise is a major undertaking. Keeping chickens for home production is not an activity to be taken lightly – particularly when they have to be looked after every day. Careful thought, research and preparation are necessary. First, however, it is appropriate to establish what is meant by free-range, and to address some other key questions.

1 KEY QUESTIONS

'No question is so difficult to answer as that to which
the answer is obvious.'

– George Bernard Shaw

WHAT IS FREE-RANGE?

In the past free-range was a general description, indicating only
that poultry were allowed to range over the fields. There was no
legal restriction on the number of birds to the acre if eggs from these
flocks were sold. A flock was usually confined to one field, then
moved to another, in sequence, as the grazing became exhausted.
It was a familiar sight to see a field brown with Rhode Island Reds
or snowed under with White Wyandottes, so congested that it was
virtually impossible to detect areas of green between them. It was
common practice to fold or temporarily house chickens on newly
harvested arable fields to glean the spilled seeds and clear the
land of residual pests. They frequently followed cattle which were
moved on as the larger grasses were eaten, making new, short
growth available for the poultry. The birds' scratching activities
broke down and dispersed the cow pats, providing a useful harrow-
ing service. Poultry fitted well in a mixed farming economy, particularly as their foraging helped to control pests such as slugs and
leatherjackets, two banes of the grower – but their management at
that time shows marked differences from free-ranging as we know
it now!

Today, free-range is a specific term. European Community regulations demand that eggs offered for sale with this description must
be from flocks which are kept in the following conditions:

- The hens must have continuous daytime access to open-air runs.
- The ground to which the hens have access must be mainly
  covered with vegetation.
- The maximum stocking density should not exceed 1,000 birds

3

per hectare of ground available to them (400 birds per acre, or 1 bird to every 10 square metres).

- The interior of the building must conform to one of two standards:
  Perchery – where there is a minimum of 15 cm perch space per bird and a maximum stocking density of 25 birds per square metre of space in that part of the building available to the birds.
  Deep litter – where at least one-third of the floor area should be covered with litter material such as straw, wood shavings, sand or turf, and a sufficiently large part of the floor area available to hens is used for the collection of bird droppings. The stocking density with this type of house should not exceed 7 birds per square metre of available floor space.

Where eggs are sold with the description 'semi-intensive' the hens must have the same conditions as those referred to above, the only difference being that a maximum stocking density of 4,000 hens per hectare of ground (1 hen per 2.5 square metres) is allowed.

Eggs sold under the description 'deep litter eggs' must be from hens kept solely under the conditions described for a deep litter house, while those with the label 'perchery eggs (barn eggs)' must be from those kept in conditions such as those outlined above for the perchery house.

From these definitions, it will be seen that the only genuinely free-range operation is that which complies with the first set of conditions. If a free-range operation is being considered, then let it be just that!

The regulations ensure that abuses such as keeping a flock indoors for most of the day and allowing them out briefly into a mud-filled patch in order to use the description free-range do not occur. If they do, the culprits can be prosecuted under the Trades Description Act.

It is important to emphasise that the regulations refer to maximum stocking rates, and that these may be inappropriate for certain conditions. For example, if the land is not naturally free-draining it will be necessary to reduce the number of birds per hectare, otherwise quagmire conditions may result. The prevalence of certain diseases in localised areas may also necessitate a reduction in stocking density; it is a good idea to obtain veterinary advice on this question so that problems can be avoided.

The ideal, as many small producers have found, is a maximum of 100 birds to the acre, with the total number of birds divided

4

into small, separated flocks of this size. This not only reduces the risk of disease, but if the flocks are of different age groups there is less likelihood of a seasonal dip in egg numbers. There may be producers who would find this low stocking density unacceptable in relation to the relative cost of land. This is an important factor and one which needs to be taken into consideration when costing out an enterprise. It could be that a stocking density of 200–300 birds per acre is satisfactory, as long as the nature of the land is suitable. It is here that obtaining expert local advice from veterinary sources and from the local Agricultural Development Advisory Service (ADAS) office is recommended. The local telephone directory will give the relevant address and telephone number.

As far as table poultry is concerned, those which are described as free-range should also have access to grazing, although it is recognised that they will need to spend the last week or so indoors, during the so-called finishing period.

ARE THERE ANY LOCAL BY-LAWS WHICH APPLY?

This may seem an obvious point to check, but it is surprising how frequently it is forgotten that there are by-laws which prohibit the keeping of poultry in some localities bordering on urban areas. There may also be regulations prohibiting the setting up of a farm shop because of potential traffic hazards. There was a recent case of a producer who was given an enterprise award by one government department, only to be closed down by another because he was too successful in attracting traffic to his site! The right hand of the establishment does not necessarily know what the left hand is doing, but checking with the local authority is a sensible first step whatever the location.

ARE THERE ANY RESTRICTIONS RELATING TO THE PROPERTY?

It is advisable to check title deeds, tenancy agreements or lease contracts which apply to the property in case there are restrictive covenants relating to livestock. It would not be sensible to invest time and money in housing, stock and equipment, only to find that there are clauses which prohibit chickens on the site.

IS PLANNING PERMISSION REQUIRED?

It depends on the scale and nature of operations whether local authority planning permission is required. If the enterprise is fairly small, using existing or moveable buildings and selling eggs at the farm gate, it is unlikely that planning permission will be needed. If new buildings are erected or if a farm shop is started, either as a new building or as an extensive renovation of an existing one, then change of use planning permission is likely to be needed. Bear in mind the advice given above about not causing a traffic hazard with a farm shop, badly sited signs or lack of parking and turning facilities. Contact the local authority before embarking on a project and ask for their advice. They may be able to offer information which is particularly useful and relevant to the local situation.

IS IT NECESSARY TO BE REGISTERED?

If you sell eggs (even from one bird) you are required to register your flock with the local authority. There is also a requirement to test your birds for salmonella on a regular basis if you are selling eggs, or if you have more than 24 birds in a breeding flock. The legislation which covers these requirements is the Poultry Breeding Flocks and Hatcheries (Registration and Testing) Order, 1989, and the Poultry Laying Flocks (Testing and Registration) Order, 1989. Guidance notes are available from your local animal health office which is listed in the telephone directory under Agriculture, Ministry of.

WHAT SORT OF SITE IS SUITABLE?

This is an important question, for the *type* of land is of key importance to free-ranging birds. The ideal is a light, free-draining but fertile soil, capable of producing a healthy sward of short grasses. Any badly drained, boggy land is a haven for parasites such as flukes and coccidia. Hens scratching about in badly drained soils will produce boggy conditions relatively quickly. Chickens also require a sheltered location, rather than an open prairie with lashing winds.

If produce is to be sold direct to the public from the site, it needs to be in an accessible place, not down an isolated track.

ARE MAINS SERVICES ESSENTIAL?

The provision of a clean drinking water supply is essential and if a mains supply is not available the source should be tested for purity. Where the water is for agricultural purposes, the water authority may deem it necessary to provide a metered supply so that the producer pays for the quantity used.

Electricity is required for the provision of cool storage for the produce, as well as for supplementary lighting in the poultry house during winter.

If eggs are to be sold from the farm gate, access via a made-up and maintained road is needed. Cars are unlikely to call otherwise. Where a distributor is used, there need to be proper access and turning facilities for regular collection.

If the unit is a domestic one and no produce is being sold, then obviously the above does not necessarily apply.

WHAT SORT OF BUILDINGS ARE REQUIRED?

The type of buildings used will depend upon the scale of operations. There are basically two options – moveable houses and static ones, the latter tending to be used by larger enterprises. Within these two categories there is a wide choice, including the use of existing and refurbished buildings, purpose-built units and houses made of different materials such as wood, brick and polythene. Further details are given in Chapter 2.

WHAT SORT OF BIRDS?

If chickens are to be kept for egg production, it is best to choose a modern hybrid strain selectively bred for this purpose. A hybrid will eat less and lay more eggs than the traditional breeds, and adapt well to outside conditions. The exception is where very dark brown, speckled eggs are required. In this case, it is worth considering an old breed such as the Maran.

If table birds are to be produced, one of the new, slow growing strains of red- or black-feathered hybrids is a good choice as they have been selectively bred for the free-ranging sector. Alternatively, an old breed such as the Light Sussex is worth considering.

For the household unit, the choice is not as crucial, and may include any of the traditional breeds – whichever happens to be popular. Small, specialist breeders will also tend to concentrate on the breeds in which they have a particular interest. (See Chapter 4 where this question is discussed in detail.)

IS THERE A DEMAND FOR FREE-RANGE PRODUCE?

Free-range eggs are said to be popular with consumers. How true this is in a particular locality will only emerge as a result of careful market research. There is really no alternative to going out and doing some research. Talking to local shops, hotels and butchers is the obvious point at which to start, and they will need to be assured of clean, fresh produce on a regular basis. Few people are interested in a summer glut followed by the drying-up of supplies over the winter.

On the general question of demand, indications are that a substantial minority of consumers are prepared to pay a premium price for free-range eggs. At the time of writing this was approximately 35p per dozen above the price of other eggs, but this figure should be taken as a general indication only, for both demand and prices fluctuate, sometimes quite dramatically.[1]

As far as non-intensive table poultry is concerned, the whole-sale price (live weight per pound, ex-farm) was approximately double that of intensively produced broilers at the time of writing, and organically produced birds which met the requirements of the United Kingdom Register of Organic Food Standards (UKROFS) were producing 3–4 times the return on intensive birds in the south-east of England.[2] Predictions have been made that free-range production will eventually account for 30% of the total market.[3]

All the indications are that there is a considerable and growing demand for both free-range eggs and table birds, but situations can and do vary, so anyone considering such an enterprise is advised to carry out careful, individual research. ADAS offers a consultancy service which includes market research for specific localities.

IS THERE AN OPTIMUM SCALE OF OPERATIONS?

There is an optimum scale for a large unit, for a medium sized unit and for a small unit. What is important is to establish which is the right one for you. What may suit one person may not be right for another.

As a general rule, the larger the operation, the higher the setting-up costs but the lower the subsequent running costs. For example, a large producer will have the benefit of reduced prices for bulk buying or feedstuffs, particularly if the feeds are mixed on the farm. Where there is automatic egg collection, the labour costs will also be lower than on a small enterprise which is usually more labour-intensive.

There are, of course, people who keep poultry purely for pleasure, or who may only be interested in selling free-range produce on a small, part-time and local scale. They represent the majority of producers and have an important role in rural life. Their importance should not be underestimated or dismissed with the disparaging comment that they are 'only hobbyists'. There are many excellently run small units, often combined with farm-gate sales of other home-grown produce. In such cases the cost factors may not be as crucial as they are for larger producers, but there is a lot to be said for not having 'all one's eggs in one basket'!

Some producers prefer to start in a small way anyway, and gradually expand as they see the possibilities of the market. There is a great deal to be said for this approach, particularly as it generally does away with the need to borrow large sums of money.

WHAT SORT OF OUTPUT CAN BE EXPECTED FROM A FREE-RANGE UNIT?

There has always been a fund of information about egg output from batteries. The breeding companies issue regular statistics based on 'hen-housed averages' (HHA) for their particular strains of birds. A hen-housed average is calculated by taking the total number of eggs produced during a 52-week laying period, and dividing this by the total number of birds in the original flock (including any subsequent mortalities). If, for example, 99 birds from an original flock of 100 produced 28,236 eggs in a 52-week period, HHA is 28,236 divided by 100, which is 282. This would be an excellent average, with 260 being a more likely figure.

Statistical information about free-range systems has not been available until comparatively recently, mainly due to the dearth of such enterprises. To someone considering starting a free-range business, there is an understandable worry that the number of eggs produced may be far below what is possible with a battery system. Recent research conducted by the West of Scotland Agricultural College has produced results which should go some way towards dispelling such worries. In their study, 38 laying hens of each of 2 strains, Shaver 288 and Warren SSL (the latter since renamed ISA Browns), were kept on free-range and compared with 186 of each strain housed in battery cages. Each strain was reared inside on the floor until 18 weeks of age, then transferred to its respective environment. Egg production of the free-range birds initially lagged behind but subsequently caught up and overtook the battery birds. The results are indicated in Table 1.1. The time gap between 68 weeks, when the study was paused, and 72 weeks, when it was resumed, represents the moulting period.

Table 1.1 Mean rate of lay percentages to 68 weeks

| | To 68 weeks | | 72 to 100 weeks | |
	Caged	Free-range	Caged	Free-range
Shaver	79.4	78.3	70.6	72.1
Warren	76.6	76.5	67.6	72.1

Source: B. O. Hughes and P. Dun, *A Comparison of Two Laying Strains: Housed Intensively in Cages and Outside on Free Range*, West of Scotland Agricultural College, Research and Development Publication No. 16, 1988.

Other factors which emerged from this study were that in both strains, measured feed intake and mean egg weight were greater in the free-range hens. They supplemented their diet to an unknown extent by eating grass, but average body weight and plumage condition were superior to those of the caged birds. The researchers do stress, however, the importance of a high standard of management, precise control and the need to have small flocks well below the official maximum stocking density.

The National Agricultural Centre Poultry Unit at Stoneleigh has also carried out comparative studies on free-range and caged birds. Their initial study indicated that egg production for the average free-range layer was 266 compared with 271 for the average caged bird.[4]

Data in Table 1.2 shows the production records of their free-range flocks over a period of 3 years. The second flock experienced a high mortality rate, including fox losses, but subsequent experience did not see a repetition of this after electric fencing was installed. Despite the fluctuations in market price, the premium on free-range eggs meant that they still produced an overall profit, particularly as they were able to sell ungraded eggs in their farm shop.

Table 1.2 Production records of free-range flocks over a period of 3 years

Period in weeks	1st flock 20–72	2nd flock 20–68	3rd flock 20–68
Number of birds	304	600	600
Average eggs per bird	285.5	239.6	182.5
Mortality	7.24%	32.2%	9.0%
Egg price per dozen	73.9p	59.9p	54.0p
Feed cost per dozen	28.7p	28.6p	*
Margin eggs over feed per dozen eggs sold	46.1p	31.3p	28.3p

* Feed prices fluctuated considerably, but fell overall.
Source: NAC Poultry Unit, *Progress Reports*, 1986–88.

The Ministry of Agriculture has also produced information on production levels for free-range flocks. Details of their comparison of caged, perchery/barn and free-range birds are shown in Table 1.3.

Table 1.3 Comparison of caged, perchery/barn and free-range flocks

	Caged	Perchery/barn	Free-range
Number of eggs	276	264	260
Mortality	5%	5%	8%

Source: *Alternative Systems of Egg Production*, MAFF/ADAS, 1986.

These various findings indicate that a free-range system can produce almost as many eggs as a battery system, and in some cases the output has subsequently been better than that of batteries, particularly in the production of large eggs. This does, of course, depend on a combination of good stock, good conditions and good management.

There is little point in comparing non-intensive table bird and intensive broiler output, for the criteria are different. Broilers have been bred for rapid development and large growth in as short a time as possible. Non-intensive table birds have been bred for slower growth, with the aim of producing a quality product in terms of flavour, texture and freedom from additives.

WHAT SORT OF COSTS AND MARGINS ARE INVOLVED?

The main setting-up costs of a free-range enterprise, assuming that the land is already available and suitable, are perimeter fencing, housing and the stock itself. As far as running costs are concerned, the main cost is that of feed which is approximately 65% of the total running costs. Feed consumption will vary from around 130 g per bird, per day, in summer, to 150 g or even higher in winter, although the aim should be to keep this as low as possible by ensuring that housing is adequately insulated.

It is difficult to give precise figures of costings and margins, for such information dates fairly quickly and much depends on the type and scale of the enterprise. The following material indicates some of the available data, but it should be regarded with caution because situations and factors change rapidly, and there is often a difference between geographical areas. The main interest is in the relationship between costs and margins, and how they are worked out, rather than in the figures themselves.

MAFF/ADAS gives the following findings on the relative costs of producing a dozen eggs, excluding the cost of land.

Table 1.4   Cost of producing a dozen eggs

	Pence
Feed	34.1
Livestock depreciation	7.9
Deadstock depreciation	7.1
Labour	13.8*
Electricity	0.9
Medication	0.3
Miscellaneous	1.3
Total	65.3

* Automatic egg collection would halve this cost.
Source: Alternative Systems of Egg Production, MAFF/ADAS, 1986.

12

The Ministry also suggests that the capital cost, including a new house but excluding land, is about £12 per bird, with an expected income of around 75p per dozen eggs. The type of house used and the scale of the enterprise will obviously have a considerable bearing on this. Those who are utilising existing or second-hand buildings will have much lower setting-up costs, although it does not pay to skimp on the provision of effective electric fencing. Polythene housing, for example, is currently around £4 per bird, while second-hand buildings are about £2.50 per bird.

It is also difficult to be precise about the likely return on free-range eggs. Although the market sources referred to earlier do give useful information about current prices, they are inevitably generalised, average prices. In February 1989, for example, at a time when a salmonella scare had caused a large drop in overall egg sales in Britain, my local supermarket was selling size 1 free-range eggs at 94p a dozen. At the same time, a free-range farm shop was selling size 1 eggs at £1.40 a dozen. Its trade had shown an initial dip when the salmonella scare surfaced, but then recovered and exceeded that of the supermarket. In February 1989, *Farmers Weekly* also reported on a free-range egg producer who was currently getting the following farm-shop prices:

Size 1 –	£1.40 per dozen
Size 2 –	£1.20
Size 3 –	£1.00
Size 4 –	.80p

By the beginning of 1990, the returns were even higher, with one retail chain selling size 1 free-range eggs at £2.10 per dozen.[5]

Ultimately, it is one's own costings and returns which matter, and it is important to be able to record and forecast these. A sample budget sheet to work out your own costings and margins is shown in Table 1.5.

Housing, feed and breeding stock companies will also often supply data on the costs and returns on a free-range enterprise, with their own products, naturally enough, forming an integral part of the equation. Such data can be most useful for anyone thinking about starting an enterprise. An example is given in Table 1.6, but again it should be emphasised that the actual figures have dated since they were produced. However, the method by which costs and returns are worked out and demonstrated are of ongoing interest.

Table 1.5 Sample budget sheet (fill in your own figures)

Costs	£.p.	Income per bird	£.p.
POL pullet	——	—— eggs at —— per doz	——
Feed (—— g/bird per day)		Cull bird	——
at —— per tonne	——		
Electricity	——		
Veterinary	——		
Miscellaneous*	——		
Total	——	Total	——

* Miscellaneous includes every other cost such as water, repairs, equipment, supplies, stationery, telephone, guard dog, etc.
The profit margin (what is left after subtracting total costs from total income) should also cover items such as depreciation, interest on capital, etc.

And what of the very small producer? It is likely that a small unit will make use of an existing building and adapt it accordingly, or purchase a new house for anything from 25–300 birds. For home use, houses and runs for half a dozen birds are widely available. For relatively low numbers a moveable house is ideal, and there are various good constructions available, new or second-hand, from a wide variety of suppliers. Many suppliers now sell a complete 'package' to include house, nest boxes, feeders, drinkers and point of lay birds.

The types of costs involved are indicated in Table 1.7 and the advice given on how to fill in your own budget sheet, such as that detailed in Table 1.5, is also highly relevant.

The setting-up costs outlined in Table 1.7 would be considerably reduced if second-hand housing and equipment were utilised. At the time of writing, a suitable second-hand house for around 100 birds costs in the region of £200, while second-hand equipment is from one-third to half the price of new, depending on condition. Farm sales or private sales advertised in the small farming press and in local newspapers are a fruitful source of supplies. It should be emphasised that any second-hand housing and equipment needs thorough cleaning and disinfecting before stock is introduced.

An example of the costs of a small domestic unit – a 6-bird flock for family egg supplies – is detailed in Table 1.8. Again, costs would be reduced if second-hand or home-made housing were utilised. If winter eggs are to be guaranteed, an investment would need to be made for a supplementary lighting system.

Table 1.6 Costs and returns for a large free-range unit

Assumption: housing at 18 weeks, depleted at 72 weeks, 4 week turnaround.
Value of manure set against removal.

Income		Per bird (£)
Assume 253 eggs (22 dozen per month for 11.5 months)		
Month 1	nil	
Month 2	22 eggs at 35p per dozen	0.64.2
Month 3–12.5	22 eggs at 70p per dozen	13.47.5
		14.12
Cull birds 2800 @ 5lb @ 12p per lb = 60p per bird		0.60
	Income per bird	14.72
(If price realised is 80p per dozen)		+ 1.92
		(16.64)
(If 23 eggs per month @ 70p per dozen)		+ 0.63
		£17.27

Expenditure (with % before capital costs)		(£)
Feed 132 g per day @ £145 per tonne	(61.4%)	7.23
Birds, including vaccination	(18.8%)	2.22
Labour (1 man-year – 4,000 birds		
+ 8 hours overtime @ £132 per week	(16.2%)	1.91
Administration and miscellaneous	(1.7%)	0.20
Electricity		0.07
Repairs and maintenance		0.05
Insurance		0.03
Medication/vet		0.03
Water		0.03
		11.77
(If feed cost is £150 per tonne)		+ 0.25
		£12.02

Therefore margin per bird before capital investment			
Eggs per bird		253	265
With feed @ £145 per tonne		2.95	3.58
If price 80p per dozen		4.87	5.50
With feed @ £150 per tonne		2.70	3.30
If price 80p per dozen		4.62	5.25

Capital costs (for 4,000 free range birds, excluding cost of land and fencing)
(cost per bird)

Birds per square metre	9.1
Erected building	5.17
Feeders/drinkers	0.77
Nest boxes (rollaway)	0.95
ACNV control and fans	0.19
Bulk bin	0.40
Electrical plumbing	0.40
Site works (full floor concreted)	1.09
Total	£8.97
For 4,000 hens	£35,880.00
(This excludes the cost of land and fencing)	

Source: Challow Buildings Limited, European Poultry Fair, Stoneleigh, 1987.

15

Table 1.7 Costs of setting up a small 100 bird unit (land excluded)

Item	Cost (£)
Moveable house (new)	450
Feeders	40
Drinkers	40
Perimeter fencing	165
Internal electric fencing	125
Point of lay birds at £3.50 each*	350
Feed storage (using dustbins)	30
Light system and time switch	150
Egg grader (optional)	75
Total	£1,425

* This is an overestimate; birds are usually cheaper depending on the quantity.
Source: author's costings, 1990.

Table 1.8 Costs of setting up a 6-bird free-range unit

Item	Cost (£)
Moveable house or house and run (new)	100
Feeder	12
Drinker	12
Point of lay birds at £3.50 each	21
Feed storage (using dustbin)	12
Total	£157

Source: author's costings, 1990.

IS INCOME TAX PAYABLE?

All sources of income must be declared on a tax return for the Inland Revenue. Everyone has basic allowances on which income tax is not levied but, above this, the standard rate of tax will apply. An accountant will advise on all aspects of this question and prepare your accounts for submission to the Inland Revenue.

IS IT NECESSARY TO BE REGISTERED FOR VAT?

It is necessary to be registered for VAT if your turnover is likely to be in excess of £25,000 a year (1990 figure). There is currently no VAT on foodstuffs, so you will not be charging this tax on eggs or table poultry but, if registered, you will be able to claim back all VAT on items you purchase that are relevant to the business. For example, stationery, egg boxes and services all carry VAT. Keep a careful record of these and then claim on your VAT return. If you are not certain which items carry VAT, enquire at the VAT department of your local Inland Revenue office. The address is in the local telephone directory.

In a business which is dealing with zero-rated products such as foodstuffs, the producer is usually a net beneficiary from the VAT system.

ARE BUSINESS RATES PAYABLE?

Every person over 18 (apart from special cases) pays poll tax, with the rate being set by individual local authorities. A business is liable for a business rate, but this will depend on its scale and nature. Farms do not pay rates, but a farm shop is likely to do so, particularly if there is a substantial car park. Small farm-gate sales to neighbours are unlikely to attract business rates, but the whole area is somewhat grey, with different local authorities making different decisions. The local branch of the National Farmers' Union (NFU) is often a good source of help and advice on such questions.

IS AN ACCOUNTANT NECESSARY?

An accountant is a professional, able not only to draw up and present your financial records to the Inland Revenue, but also in a position to offer sound advice on how to save money and possibly avoid some tax. (Avoiding tax is a legitimate activity, not to be confused with evading tax, which is unlawful.)

Unless you are a financial expert, appoint an accountant at an early stage! There are, of course, good and bad accountants. Ask around in the local business community and you will soon establish which are the effective ones and which are the ones to avoid. Accountancy companies with many employees will charge more than a small practice of one or two accountants because their overheads are higher – but going for the cheapest may not always be the best policy!

The more efficient you are at keeping your own records, the less time the accountant will need to spend on your books, and the less you will have to pay him. If, for example, you keep your records on a computer and are able to supply the accountant with a computer disc copy of your data, he can then feed it into his computer and view the figures quickly and easily. Personal computers are effective and time-saving tools, and any small business is well advised to consider using one. Most IBM compatible computers are suitable for business use, and a wide range of software programmes for word processing, keeping records and financial transactions are available. Consult your accountant for his advice on the most appropriate system and bear in mind that there are now many local computer training courses for the small business user. Enquire at your local education authority.

WHAT ABOUT BORROWING MONEY?

There are two schools of thought when it comes to borrowing money to start a business. The first is to go ahead and do it, while the second view advises starting small and gradually expanding, ploughing back the profits in order to do so. Both are valid views and only the individual is in a position to decide which to choose, or whether to adopt elements of both on the basis that 'a judicious blend of both is more perfect than either'. It goes without saying that it would be unwise to borrow money while interest rates are high.

What is essential is that a decision should only be taken after asking for and receiving professional advice. ADAS, referred to earlier, provides a consultancy service which includes costing out a specific poultry enterprise. It is worth paying for such a service, particularly as you will then have the appropriate figures and cash flow projections with which to apply for a loan. The bank manager is the obvious person to talk to about a loan, but consult your accountant for his views and advice as well.

WHAT ABOUT INSURANCE?

Insurance cover is essential in any business. Find a reliable and registered insurance broker, ideally through the local small business community grapevine, and discuss with him the whole nature of your enterprise. He will be able to suggest all sorts of aspects which you may not even have considered. Those who have farm shops, where the public has access, for example, will need a comprehensive public liability coverage against accidents and other eventualities. There are also specialist companies which offer cover on livestock and farming activities. Discussions with a good, reliable insurance broker should take place at an early stage.

WHAT OTHER SOURCES OF HELP AND INFORMATION ARE AVAILABLE?

Reading as much as possible on the subject provides an excellent theoretical background, although it is advisable to be wary of some advice in older books on free-range production, because the information may be out of date. A book such as this is up to date, but it would be a foolhardy author who claimed that reading can take the place of practical hands-on tuition. It is here that agricultural colleges and other educational bodies can be of great service.

The two basic areas where help and information are required are with practical husbandry tuition and starting and running a business. There are a number of organisations which offer advice. Some are free, while others charge for their services. Remember that all such charges can be offset against the business, so it is important to start as you mean to go on, and obtain the appropriate receipts.

The main sources of advice are listed below, but there is further information about appropriate organisations in the Reference section at the end of the book.

Agricultural Development and Advisory Service (ADAS)

This organisation offers a complete consultancy and advisory service for the free-range producer, including advice on how to set up and run a unit, and researching and reporting on individual sites and enterprises. The first consultation is free but subsequent ones

19

are charged. The address and telephone number of your nearest branch will be in the local telephone directory.

Agricultural Training Board (ATB)

This is a national organisation with local branches listed in the local telephone directory. It will organise and run a practical course in a particular area if there are enough people likely to take advantage of it. Courses are often run in conjunction with agricultural colleges, but any local group can apply to put on a specific course, as long as there are enough people interested. These courses are excellent value and an essential prerequisite to starting a free-range unit.

Agricultural colleges

A list of colleges which provide courses on free-range management is given in the Reference section at the end of the book. Many of these courses are run in conjunction with the ATB. They offer excellent training in practical management, particularly as many colleges have their own free-range units and flocks available for demonstrations. They also offer a wide range of courses which are relevant to the rural business sector, such as keeping records, using a computer, running a family farm or starting a farm shop.

Local educational organisations

Local colleges of further education frequently schedule evening or part-time courses on aspects of running a business. They are extremely valuable, particularly if they are up to date in their coverage of subjects such as the use of computers in the small business. Local schools are also the venue for useful evening courses on computing and allied topics. Enquire at the offices of your local education authority.

Freephone enterprise

Nothing could be simpler than picking up the telephone, dialling 100 and asking for Freephone enterprise. (As the name implies, there is no charge for the call.) It is a service which provides help and information for anyone thinking of starting their own business. A caller can be put in touch, for example, with the Small Firms

Service which offers up to 3 free counselling sessions about all aspects of starting a business.

Enterprise allowance scheme

This is a scheme which was set up to help unemployed people start their own business. Applicants should be over 18 but under retirement age, and should have been unemployed for at least 8 weeks. They should be receiving unemployment or supplementary benefit at the time of application. The only other condition is that the applicant should have £1,000 available, either in his own account or as a loan from the bank. If the application is successful, the scheme will pay the applicant £40 a week for a 52-week period, to help cover personal costs while the enterprise is in its difficult initial period.

Asking and pursuing all the key questions in this section of the book has hopefully provided a sound basis from which to proceed further. In the following chapters we shall look at the practical details of poultry management, before returning to the question of marketing in Chapter 10.

References

1 United Kingdom Egg Producers' Association (UKEPRA), 1990.
2 Author's research, 1990.
3 *Poultry World*, February 1990.
4 NAC Poultry Unit, Newsheet 5, 1985.
5 *Poultry World*, February 1990.

2 SYSTEMS AND HOUSING

'A system is only as good as the one who operates it.'

SYSTEMS

Fifty years ago, free-ranging chickens were normal. These days, free-ranging is regarded as an alternative to the battery system, although it must be said that there has always been a minority of producers who have adhered to traditional practices.

The basic principle of free-ranging is a simple one. The birds are provided with a house from which they range over a defined area of land that is changed at frequent intervals. The poultry-keeper with 6 chickens is operating on the same principle as the commercial enterprise with 5,000 birds; the only variations are in scale, in the type of house used and in how the land is managed.

The moveable house system

This is the system used by most small, free-range producers and has stood the test of time for many generations. The house, for anything from 6–300 birds, stands in an area of paddock to which the birds have access for a specific period of time, usually around 6 weeks. It is then moved to a new area, with the original paddock being closed to the birds and allowed time to rest and recover.

Moving the houses can be done manually or with the aid of a tractor or other vehicle. They are usually equipped with wheels, skids or carrying poles for this purpose, depending on the size and type.

Although the EC regulations set the maximum stocking density as 1,000 birds per hectare (400 per acre), many small producers have a maximum of 100 birds to the acre because there is far less likelihood of a concentration of disease-causing organisms with this level of use. Where a greater density is allowed, they argue,

A moveable house for up to 300 birds with nest boxes on either side and a central droppings pit with slats. (Challow Products Ltd)

producers may not experience problems for the first few years but, in the long term, problems will emerge. This has certainly proved to be the case since commercial free-range production reappeared in Britain. The lower stocking rate is also one which accords with the recommendations of most of the animal welfare organisations, and is the standard which I have always followed with my own poultry.

The kind of house used with this system is the perchery type – 'allowing a minimum of 15 cm perch space per bird and a maximum stocking density of 25 birds per square metre'. The perches may be parallel or stacked; the latter has them placed at different levels, but obviously not directly above one another. This arrangement makes efficient use of space, provides extra exercise within the house and allows the birds to keep each other warm in particularly cold periods. A droppings board placed under the perches enables the droppings to be cleared away at frequent and regular intervals. More information on this is given in the latter part of this chapter.

The obvious disadvantage of the moveable house is the time and

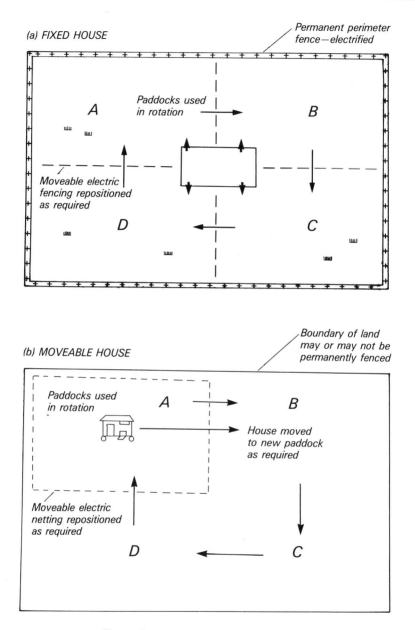

(a) FIXED HOUSE

Permanent perimeter
fence—electrified

Paddocks used
in rotation

A

B

Moveable electric
fencing repositioned
as required

D

C

(b) MOVEABLE HOUSE

Boundary of land
may or may not be
permanently fenced

Paddocks used
in rotation

A

B

House moved
to new paddock
as required

Moveable electric
netting repositioned
as required

D

C

Figure 2.1   The principles of free-range

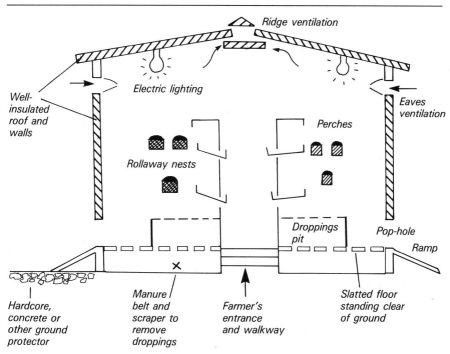

Ridge ventilation

Electric lighting

Well-insulated roof and walls

Eaves ventilation

Perches

Rollaway nests

Droppings pit

Pop-hole

Ramp

Hardcore, concrete or other ground protector

Manure belt and scraper to remove droppings

Farmer's entrance and walkway

Slatted floor standing clear of ground

Figure 2.2   End-on view of a large perchery house

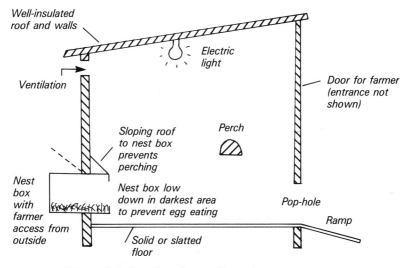

Well-insulated roof and walls

Electric light

Ventilation

Door for farmer (entrance not shown)

Sloping roof to nest box prevents perching

Perch

Nest box with farmer access from outside

Nest box low down in darkest area to prevent egg eating

Pop-hole

Ramp

Solid or slatted floor

Figure 2.3   Interior of a small perchery house

effort that has to be spent on moving it, although it is arguable that this is obviated by the lessened risk of disease.

The fixed house or static system

Large units tend to use fixed houses in an area fenced off by a permanent perimeter fence. Within this area, the ground is divided into paddocks which are made available in rotation. As the house cannot be moved, bird access to different areas of ground is controlled by having a number of pop-holes (bird exits) in different parts of the house, which are opened or kept closed, depending on the areas to be grazed. Portable electric netting is also used to control paddock access. (See Chapter 3 on Land Management.)

Fixed houses which are used for free-range production may be of the perchery or deep litter type. The latter has 'at least one-third of the floor area covered with litter material and a stocking rate not exceeding 7 birds to the square metre of floor space'.

Where eggs are sold with the description perchery/barn eggs rather than free-range, it indicates that they are from birds which are not free-ranging but confined to a house where they are able to move about and perch. There are some producers using a form of perchery system which is, in effect, a large cage allowing the birds

A large perchery house for 2,500 birds, with slatted floor over a deep droppings pit. Density is 15 birds per square metre and feeding is automatic. Rollaway nest boxes are used and there is both natural and electric ventilation. Drinkers are provided outside as well as inside.
(Mick Dennett. Bibby Agriculture)

26

to move up and down via a series of perches.

Fixed houses are more easily automated and managed than moveable ones, but they do have disadvantages. The hens have a tendency to remain close to the house, churning up the ground in the immediate vicinity and thereby creating a potential health hazard. Giving their grain ration at a distance from the house is one way of encouraging the hens to wander further afield, but it does not solve the intrinsic problem. This question is discussed more fully in Chapter 3.

Combined systems

Some producers may use a combined system. This is not an officially defined term for the purposes of egg sales, but merely a description of the type of enterprise some poultry farmers will find appropriate to their needs. For example, a farm may have its flock free-ranging in the spring, summer and autumn months, abiding by all the requirements of the free-range system, and selling eggs with this description. In winter, particularly in the colder areas of the country, weather conditions may be so extreme that it becomes necessary to house the birds for the duration, and sell the eggs under the requirements and description of perchery/barn eggs. With the advent of gentler conditions, the birds then revert to free-range status and conditions. This is a common-sense practice where the enterprise is a small-scale one, selling eggs at the farm gate, in a particularly exposed area. There is unlikely to be confusion in the minds of the customers if they can see the birds for themselves, whether in or out of a barn.

Problems arise where battery egg producers, or large perchery/barn producers have free-range units alongside their other houses. There is nothing unlawful about this, of course, but it can be confusing for the customer, and it may provide a temptation for the unscrupulous. There is no doubt that some free-range enterprises are recent innovations by a small number of battery producers, anxious to obtain the premium on free-range eggs at a time when the return on battery eggs has been dropping. Some small enterprises have also rushed into production for similar reasons. There have been prosecutions where battery eggs have been sold as free-range ones. It should also be mentioned that responsible animal welfare organisations have expressed reservations about both the permanent perchery/barn system and the battery system. Unless geographical location deems otherwise, therefore, if a free-range unit is being planned, let it be a real free-range one!

The house and covered run

Mention must be made of the covered run system which is essentially a house with a roofed poultry run extending from it. It is an attempt to provide hens with conditions where they are able to scratch, perch and take dust baths, while ensuring that they are in relatively protected conditions. Traditionally, covered straw yards were used and there has been a recent resurgence of interest in this approach.

A covered area can quite easily be incorporated into a general free-range system between the house and paddock. It is, in effect, a sheltered area between the outside world and the house. If managed properly, it can be a positive advantage in ensuring that eggs are kept clean. One of the problems that may arise with free-range birds is that they take mud into the house on their feet, and this is transferred to the eggs. The straw yard or other protected run provides an effective doormat service! An alternative is to concrete the area immediately around the pop-holes in a fixed house, or to use some kind of verandah or ramp.

An existing house can be given an extended roof on one side, with poles supporting the sides, rather like the traditional pole

A static house with large protective porch over the pophole area, and verandah to keep mud out of the house. (Onduline)

28

barn structure for turkeys. The birds are able to go in and out of the house, range onto grass areas if they wish to, or have the benefit of the sheltered space under the projecting roof. The area should be regularly cleaned and provided with litter material such as a thick layer of sand or regularly replenished wood shavings. The birds can still be encouraged to wander further afield by providing their grain ration at a distance from the house, and in a different position every day.

There are disadvantages with the covered run system, as there are with every system. The main one is that, if a lot of straw is used, it may act as an encouragement to rats. The organisms causing the disease coccidiosis may flourish if the area is allowed to become damp, although careful management should prevent this happening. It is also not unknown for birds to make nests there, rather than laying in the house nest boxes. Again, careful management is required. (The term 'straw yard' does not mean that the use of straw is essential, incidentally. It merely indicates that this was the traditional form of litter. These days, wood shavings from untreated wood are more common.)

The West of Scotland Agricultural College, in trials conducted with covered straw yards, found that there was a high output of eggs (around 80%, with a top production percentage of 92.1% between 32 and 36 weeks), but the naturally ventilated system and the relatively low temperatures resulted in a high average feed intake of 140 g per bird. Other problems encountered were floor-laid and dirty eggs, although it was stressed that sound management and husbandry techniques in the appropriate positioning of nest boxes could control these. In these trials the hens were not free-ranging, but confined to the house and straw yard.[1]

The domestic unit

It is important not to forget the small-scale poultry-keepers who represent the majority. They are usually keeping poultry for their own interest, to provide the family with eggs and possibly to sell them if there is a surplus. Such units utilise the whole range of systems, from a fixed garden shed with hens free-ranging around it to a small moveable house with attached run. Houses with covered runs, such as those detailed above, are common. In some semi-intensive units, the area underneath the house is used as an

A moveable house
with integral run.
(Walner)

A small house and
grazing unit suitable
for a garden. (Walner)

extension of the run area, a practice which makes excellent use of
confined spaces.

Some small houses are used with a permanent run. This is
not a good idea, unless the permanent run is a concrete one.
Ground which is overused can become a haven for parasites and
other disease organisms. It can also become a mud bath in winter. In such conditions poultry will sicken and die. It is infinitely
preferable to concrete the run so that it is easily hosed down. A
grassy area which becomes a quagmire does not constitute free-
range!

Hens cannot scratch about on concrete, but the provision of a
large, shallow box of clean sand will keep them happy. In summer

30

there is no reason why they cannot be put back on grass in a moveable run. It goes without saying that eggs cannot be described as free-range unless the system is fully compatible with the legal requirements.

HOUSING

Having taken a general look at the systems in use, it is now appropriate to discuss more fully the types of houses which are available and what they should provide for their inhabitants.

Domestic chickens, *Gallus domesticus*, have evolved from the jungle fowl of the tropics. *Gallus gallus*, *G. bankiva*, *G. sonneratii* and *G. lafayettii* are all present-day representatives of the wild jungle fowl group. They spend their time in their indigenous habitats scratching about in the ground litter of tropical and subtropical forests, seeking the security of tree cover at night or when danger threatens. They have no natural oil in the feathers, as do ducks, and their plumage cannot readily shed rainwater. Their domestic cousins, our familiar chickens, share similar features. In cold, damp, northern climes adequate housing is essential if they are to remain healthy and survive. Housing must not only give shelter from the cold, rain, wind and occasionally fierce sunlight, but it must also provide adequate ventilation for the birds' well-being. It should also invite the birds to lay their eggs in the nest boxes provided, as well as to feed and sleep in comfort and security.

The decision whether to use fixed or moveable houses is one which must be taken at an early stage. Reference has already been made to the fact that there are advantages and disadvantages with both systems. Much depends upon the size of undertaking, on the land and capital available, and on the attitudes of the individual poultry-keeper. Large is not always best and the budding free-range producers should make a positive effort to 'think small' before coming to a final decision.

Housing materials

There are as many different houses in use by poultry-keepers as there are variations in systems. Houses may be of stone or brick, if they are adapted for poultry from existing farm buildings. They may also be made of timber, building blocks, corrugated

31

iron, polythene or even straw bales! Some cereal farmers have made buildings out of angle iron and straw bales, effectively utilisng surplus straw following public reaction against the practice of straw burning. Some materials are better and more permanent than others, but the essential features which they must provide are:

- Adequate insulation.
- Proper ventilation.
- Protection from the elements.
- Protection from predators and vermin.
- A clean, healthy environment.

It is obvious that a building which is to satisfy all these conditions should be made of permanent materials such as brick, blocks, steel or timber, although plastic or straw bale constructions may have their place as more temporary shelters.

Insulation

The optimum temperature for a chicken is 21°C. At lower temperatures it will consume more food in order to keep itself warm. Insulating a house will not only provide a more congenial environment for the bird, but will also reduce feed costs. The greatest heat loss is through the roof, with 35% dissipating in this way, while draughts account for another 25% loss. To meet these criteria, a house roof needs to be insulated to a value of $U = 0.5 W/m/°C$ (MAFF/ADAS recommendation).

One of the most effective ways of insulating buildings is to spray the interior and exterior surfaces with polyurethane foam; 2.5 cm of foam is equivalent in 'U' value to 5 cm of fibreglass. As the foam bonds to the surface to which it is applied, rodent access is usually considerably reduced, while condensation build-up is also inhibited, as long as adequate ventilation is provided. There are specialist contractors who will undertake the job of spraying buildings in this way, ready for poultry occupation.

Buildings can also be refurbished with insulating material such as fibreglass, or with insulation boards which are widely available from DIY suppliers. Stockists should be able to confirm the 'U' value of their product to anyone who wishes to ensure that it is suitable for their building. Always wear face masks and gloves when any insulation work is being carried out.

Polyurethane foam insulation applied to the inside of a poultry house to cut down heat loss.
(British Insulations)

Steeply pitched roofs on old farm buildings may need to be fitted with false ceilings if the heat loss through the roof is high. If a false ceiling is inserted, it is important to ensure that ventilation outlets are not restricted.

As far as small buildings are concerned, insulation panels are easily tacked on to the interior of wooden houses. If the houses have been designed for complete dismantling and quick cleaning, make sure that the panels do not interfere with this facility.

A traditional and low cost practice was to tack sections of old carpet on the inside of the house, and then stuff straw between the carpet layer and the outside wooden wall. Thick polythene sheeting with insulation roll or polystyrene is a healthier and relatively cheap modern equivalent. Make sure there is adequate ventilation if this is carried out; a warning sign is when condensation is seen to be building up on the walls.

Ventilation

No bird will remain healthy for long if it is housed in a stuffy, badly ventilated house. There needs to be a continuous throughput of fresh air, balanced against the equally important need to conserve warmth. Getting the balance right is not always easy.

Large houses are best ventilated by a combination of eaves and ridge ventilation. Air comes in through the side vents and, as it warms, rises to escape through the ridge outlets above. Mechanically operated flaps can be inserted under the ridge outlet and

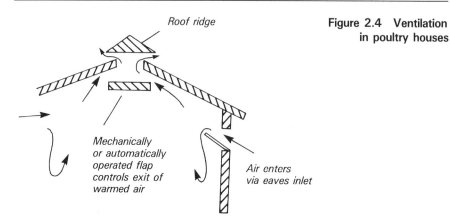

Roof ridge

Figure 2.4 Ventilation
in poultry houses

Mechanically
or automatically
operated flap
controls exit of
warmed air

Air enters
via eaves inlet

these, together with the opening and closing of the eaves, allow a considerable degree of control. Some large houses, particularly deep litter ones, utilise fans to drive the air through, but unless the house is a vast one an eaves/ridge combination should be adequate.

Small houses with relatively few birds are unlikely to need ridge ventilation. A side window panel covered with wire mesh and placed fairly high up on the wall will normally provide enough fresh air. The house should always be placed in such a way that the vent is on the sheltered side, away from the prevailing winds, otherwise draughts may ensue.

Free-range poultry house with roof ventilation and slatted verandah.

34

Adapting existing buildings

Old mushroom houses, pigsties and other farm buildings have all been used to good effect as free-range poultry housing. One enterprising farmer bought a number of old caravans which were no longer roadworthy. After stripping them out and providing nest boxes and perches, he used them as moveable houses for a number of small free-range flocks. Moving them around was comparatively easy with the aid of a tractor, as they were already equipped with towbars.

The key factors in the utilisation of existing structures is that there should be an adequate level of insulation and ventilation, as referred to earlier. Old buildings are notoriously draughty. Floors, doors and windows should all be checked to make sure that there are no access points for rats. The base of doors is a particularly common entrance point. If this area is worn it is a good idea to reinforce it with a metal panel, as shown in Figure 2.5.

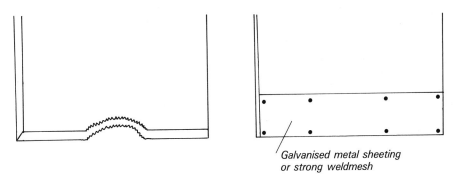

Galvanised metal sheeting
or strong weldmesh

Figure 2.5   Reinforcing worn areas against vermin entrance

Roofs

A roof must exclude rain and have enough insulation to conserve warmth. As we have seen from the previous section, it can also play an important role in ventilation.

In large houses, aluminium sheeting with internal fibreglass insulation is common as a roofing material. A synthetic material such as the plastic Onduline is also widely used, on both new houses and traditional farm buildings which are being renovated for poultry. Slates, tiles and other traditional materials are suitable, as long as

Large house roofed with Onduline and equipped with side ventilators.
(Onduline)

they are sound and do not let in water. Roofs such as these also require internal insulation and the spray-on polyurethane insulation referred to earlier is ideal.

Smaller houses may have roofs of tarred felt or corrugated and galvanised iron sheeting, materials which are quite suitable as long as they are watertight and have internal insulation. Galvanised iron sheeting on its own is very cold in winter and heats up too much in hot summers. It also has a tendency to rust at the joining points, allowing drips of water to enter.

Floors

The type of floor will depend on the type of house. A permament deep litter house will usually have a concrete floor which is proof against the burrowing activities of rats, while providing a suitable surface for litter material such as sand or wood shavings. Polythene houses are normally erected on the earth floor, without any site preparation, as they are moveable structures.

A perchery house floor ideally stands clear of the ground, preventing rodent access and, where slatted floors are incorporated, enabling the droppings to fall through. Slats which are 3 cm × 3 cm with 3 cm gaps are suitable.

Slatted floors in large houses normally have a belt or scraper system to clear away droppings. This situation will not arise in the case of a moveable house with a slatted floor for the house is merely moved on, allowing the droppings to be cleared from the site.

Many small, moveable perchery houses have solid floors and these are much warmer than the ones with slatted floors. The use of a droppings board helps to confine the droppings to the area underneath the perches, making cleaning a relatively easy task. Many modern, small houses are also designed to be completely dismantled for cleaning in a matter of minutes.

Polythene housing

A form of low cost housing which is popular with some producers

Checking the performance of a Hisex Brown flock housed in a polythene house with free-standing nest boxes and suspended feeders and drinkers.
(Plumpton Agricultural College)

is the polythene tunnel house. The price compares favourably with other forms of housing and it is a moveable structure, but there are disadvantages. The polythene needs to be a double thickness, with insulation between the layers, otherwise it is too cold. Ordinary polythene lets in too much light, a situation which can lead to pecking and cannibalism, although using black polythene counteracts these problems. A frequently used system of polythene housing utilises 2 sheets of black polythene, 175 microns thick, with 10 cm of fibreglass insulation sandwiched between them. The whole structure is then covered with another sheet of black polythene which is periodically replaced.

Polythene obviously has a more limited life than other materials: it can tear, and the effect of the sun is destructive. The use of polythene with an ultra-violet inhibitor will extend its life, however, and it does represent a considerable saving over other forms of housing.

The West of Scotland Agricultural College has carried out a research project to assess the viability of polythene free-range housing. The house was composed of a double layer of white polythene, stretched over a galvanised steel framework, with an insulating air pocket between the two layers. It was placed directly on an earth floor so that no site preparation was necessary, and it was easily moveable to a new site when required. Six hundred 17-week-old ISA Browns were introduced to the house, with the range area being divided into two so that the birds could be released alternately on to one side and then the other. The stocking rate at any one time was 770 birds per hectare, considerably below the EC regulations requirement of 1,000 birds per hectare. The birds were fed a 17% layer's mash from a mechanical chain feeder 5 times a day, resulting in a mean figure over the year of 137 g per bird per day. Eggs were collected 3 times a day. A summary of the results from 20–72 weeks of age is given in Table 2.1.

Table 2.1   Egg production in polythene housing

Hen-housed average	296 eggs
Floor eggs	2.68%
Mean egg weight (from 33–62 weeks)	62.5 g
Percentage eggs graded 1, 2 and 3	40.8%

Source: Linda Keeling and Peter Dun, *Polythene Housing for Free-Range Layers: Bird Performance and Behaviour*, Scottish Agricultural Colleges Research and Development Note No. 41, February 1988.

Egg production was good, with a hen-housed average of 296 eggs per bird, but egg sizes were low. This was thought to be due to the birds coming into lay early because of the considerable light entering the house. Black polythene would probably have prevented this. It is important not to encourage laying until the bodily frame and eating capacity of the birds are sufficiently developed for large egg production.

FITTINGS

Every house, whatever its size or type, will need to be furnished with the appropriate fittings to make it a habitable domain for the birds. The fittings include nesting boxes, perches, pop-holes, feeders and drinkers and, ideally, lighting.

Nesting boxes

The cleanest eggs are undoubtedly produced in houses which utilise 'rollaway' nests. There is no great mystique about such a construction: it is merely a series of nest boxes with floors which slope downwards, away from the hen's entrance. The nest boxes are

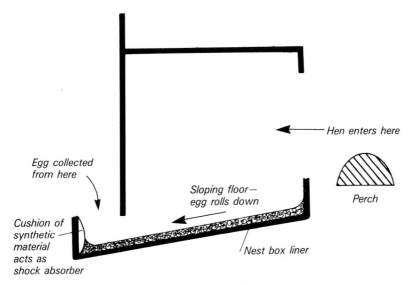

Figure 2.6   Details of a rollaway nest box

39

either lined with artificial nesting material, or have a strip of the material at the collecting side acting as a cushion. After the egg is laid it rolls backwards to the collection point, landing against the protective cushion. The collection area projects further out than the back of the nest box so that the hen cannot reach the egg. This helps to reduce the incidence of egg eating, a troublesome condition which can arise in even the best-run units. If the nesting boxes are placed in such a way that they back onto a central passageway, it is an easy task for the farmer to collect the eggs.

Where automatic egg collection is not available, it is more convenient to have access to nest boxes from the outside. (Harper Adams Agricultural College)

Free-standing nest boxes with perch access. The roof is sloped and 'fenced' to prevent perching there. (Spirofeed)

Smaller houses are normally equipped with the traditional static nesting boxes. They are usually placed so that the eggs can be collected from the outside of the house. This involves having a hinged, waterproof lid above the nesting area. Placing the nest boxes in a darkened area of the house is an advantage in that there is less likelihood of egg eating. Another technique is to suspend vertically slit plastic curtains over the nest box entrance. This does not deter the hen from entering, but it does ensure that the area is dark and private. It should be emphasised, however, that the only certain

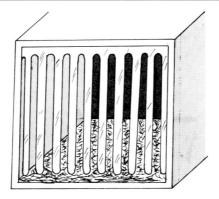

Figure 2.7 Static nest box fitted with plastic curtain

way of avoiding the problem, and ensuring the highest incidence of clean eggs, is to use the 'rollaway' nest boxes.

As a general rule, traditional nest boxes are around 30 cm high × 30 cm deep × 25 cm wide. A panel 8 cm wide along the bottom of the entrance will ensure that nesting material does not fall out. There should be 1 nest box for every 5 birds in relatively large units. Smaller units will find that 1 nest box to every 4 layers is a better ratio, while a small domestic house for 6–12 birds functions best with 1 nest box to every 3 birds.

Wood shavings make good nesting material, as long as they are clean, dry and free of dust. It is important to obtain untreated shavings from dealers specialising in livestock supplies. In the past, there have been incidences of poisoning where shavings from treated wood were used.

Some people have used sawdust as a nesting material, but this tends to clog and is not really satisfactory. Chopped straw is sometimes used, as long as it is dry and free of harvest mites. The same applies to hay, although I would not recommend it. If any of these materials become damp, mould may develop. The spores of the mould *Aspergillus fumigatus* are particularly dangerous: if these are breathed in, the lungs can be adversely affected, leading to congestion and wheezing, a condition referred to as aspergillosis. (The common name of farmer's lung indicates that it is a condition which can also affect man.)

Synthetic materials are also used in nest boxes. For example, there is a plastic material which is formed into tussocks resembling grass. It is available in rolls and can be cut to fit the nest box, with the great advantage over more traditional nesting materials that it can be easily removed and cleaned.

Perches

The hen is a perching bird, although pullets which have been cage reared often have to be shown how to perch by being placed in position. For this reason, it is best to buy pullets which have been floor reared.

The perching mechanism of the bird's foot is an extraordinarily efficient one. As the toes clamp onto the perch a type of automatic lock comes into force, enabling the chicken to go to sleep without falling off. The ideal width of a perch is around 8 cm before the top is rounded off. Once it is bevelled so that there is a rounded surface for the foot to grasp, the width is nearer 5 cm. A depth of the same size is adequate, allowing the perch to fit into prepared wall sockets at each end. These should be made in such a way that the perches can be easily lifted out for cleaning. In a small house the perch should not be more than 60 cm above the floor, otherwise damage to the feet may result from continual jumping down. Small abrasions can lead to infections and abscesses of the feet, a condition commonly referred to as bumblefoot. Details of this are given in Chapter 11 on Health.

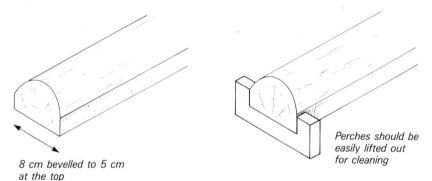

8 cm bevelled to 5 cm at the top

Perches should be easily lifted out for cleaning

Figure 2.8   Single perch

Where several perches are used, the distance between them needs to be around 30–40 cm, depending on whether they are parallel or stepped. With stepped perches, the height above the floor will vary, but the bottom one should not be more than 60 cm above

42

Hisex Browns inside the perchery house on 'A' frame perches which they were trained to use in the rearing unit. (Mick Dennett. Bibby Agriculture)

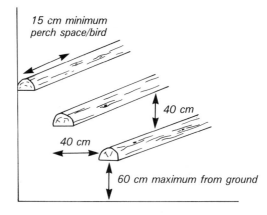

15 cm minimum perch space/bird

40 cm

40 cm

60 cm maximum from ground

Figure 2.9   Stepped perches

43

the nearest surface. Where single perches are used in relation to a droppings board, the height is normally about 20 cm above the board. With stepped perches the height will obviously vary.

Allow a minimum perching space of 15 cm per bird, to comply with the EC regulations, although many poultry-keepers would increase this to double the space on humanitarian grounds. Bear in mind that allowing more space than this will effectively reduce the birds' ability to keep each other warm in colder areas of the country.

Pop-holes

A pop-hole is a small entrance/exit through which the chickens go out of or come into the house. It is quite separate from the main door of the house, which is there for the farmer's use.

The principle of the pop-hole is that it can be opened from the outside and used to control access to a given area. In a large house, for example, several pop-holes can be installed so that access to one or more areas of paddock can be shut off by keeping the appropriate exits closed, while others are opened. Some very large houses have automatically opening and closing pop-holes, but there is a disadvantage with this system. If the farmer does not have to open the doors himself, he will be unable to check over the flock as they

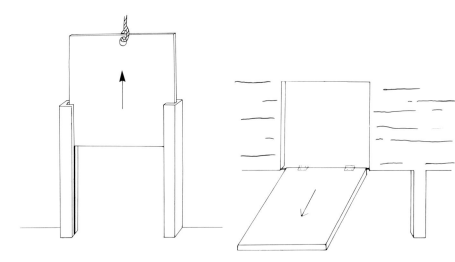

Figure 2.10 Lift-up pop-hole entrance   Figure 2.11 Drop-down ramp pop-hole

emerge and incipient health problems may not be detected until too late.

In a perchery house with a raised floor, the pop-hole will need an exit ramp so that the birds do not have to jump down or flap their way upwards. The simplest way of providing this is to have a door which lifts down from the top and becomes the ramp. Alternatively, the ramp can be a permanent fixture, with a door which slides up and down for opening and closing.

There is no need for a ramp where a house with a concrete or earth floor is used, but with both types of house it is a good idea to have the area immediately outside the pop-hole treated to avoid muddy conditions. This could be by the provision of a hardcore or concreted area, or by incorporating a protected run as discussed earlier in this chapter. In particularly exposed situations, some kind of protection will help to cut down the draughts that can otherwise contribute to reduced temperatures. Constructing a porch protection is an effective way of doing this.

Where large numbers of birds are involved, the size of the pop-hole is important. Many poultry farmers have discovered that a wide aperture allows more birds through at a time, preventing a traffic jam which may lead to the bullying of individual birds.

Raised pop-holes keep the house warm, while the ramps ensure clean feet. The exits are wide enough not to restrict movement, and this prevents bullying. (Mick Dennett. Bibby Agriculture)

Inside the large perchery house the feeders and drinkers are placed near tiered perches but above the slatted droppings area. (Mick Dennett. Bibby Agriculture)

Feeders

For small houses suspended feeders are the most appropriate, with one catering for up to 25 birds. The feeders hold the basic layer's ration – either in meal or pellet form. On a larger scale, an automatic chain feeder system is commonly used, with a spacing allowance of 10 cm per bird. Here, a layer's ration in powder form is used to prevent wastage, while a silo storage system connects with the chain feeder.

As far as the daily grain ration is concerned, reference has already been made to the practice of outside feeding in order to encourage wider dispersal of the flock. There are two possibilities: either the grain is deposited on the ground for the birds to forage, or it is placed in purpose-made feeders. A variety of such feeders is available, including some which are accessible only to chickens, not to any lighter wild birds or rodents. The principle on which such a feeder operates is that, as the chicken steps onto the pedestal to gain access to the grain, its weight has the effect of opening the feeder.

46

Wild birds, whose weight is insufficient to displace the pedestal, cannot then feed at the producer's expense.

Feed storage

All feedstuffs should be stored in dry, rodent-proof containers. The large unit will tend to use a silo-type hopper from which the feed is automatically channelled to the feeders. Smaller farms usually rely on purpose-made feed bins with tightly fitting lids, or utilise dustbins. The latter provide excellent, low cost storage and the only other requirement is a good scoop for extracting the feed.

Drinkers

Large units will tend to use an automatic system of supplying drinking water, the commonest type having nipple drinkers with

Bulk bin and auger providing feed to the poultry house. (Mick Dennett. Bibby Agriculture)

drip troughs or drip cups. A nipple drinker should be made available for every 5 birds. Suspended plastic drinkers can also be used as part of an automatic system: one drinker for every 100 birds will cater for their needs.

Suspended drinkers which are filled manually are often used by smaller units. They can be made of plastic or galvanised metal. There are also drinkers which are of the non-suspending type, but these are not recommended because litter can too easily find its way into the water.

Water will either come directly from the mains, or via a moveable tank or trough. The latter can also be used to provide water further from the house, again to encourage wider ranging of the birds.

On a domestic scale, drinkers are usually filled manually, every day. Whatever the scale, though, the importance of regular cleaning of the containers with plenty of hot water cannot be overstressed. Where an automatic system is used, it is important to follow the manufacturer's advice on cleaning procedures.

Lighting

The egg laying facility is intimately related to the amount and intensity of light which is available. Once natural light begins to decline, from summer onwards, natural light will need to be supplemented by artificial light.

The use of 40 watt tungsten bulbs or 6–8 watt fluorescent bulbs or tubes is effective, with one bulb or tube catering for up to 100 birds. The light sources are placed 3 m apart and then tested for intensity with a light meter. Light intensity is measured in lux units, with a satisfactory level being around 10 lux.

The circuit should include a time switch, so that the availability of artificial light can be programmed in advance. It should be sufficiently sensitive to allow for adjustments of a quarter of an hour and should be checked on a daily basis. A dimming facility is useful, not only to warn the birds that 'lights out' is approaching, so that they have time to perch, but also to reduce light intensity if there are problems such as cannibalism.

Specialist poultry suppliers sell a range of lighting systems, from the large scale to the small. The small poultry-keeper may wish to make his own system. An ordinary 25 watt bulb is quite satisfactory, if mains electricity is available. Where it is not, a car battery with 12 volt car bulbs can be used. Traditionally, paraffin lamps were used, but they are no longer recommended because of the risks of fire and asphyxiation.

Details of lighting systems in use are to be found in Chapter 6 on Management.

Litter

A wide range of litter material is used in houses, including straw, wood shavings, sand and shredded paper. Wood shavings are perhaps the most commonly used.

Litter should be regularly checked to make sure there are no patches of damp, and that there is no incidence of ammonia build-up. Both these situations can cause problems. Coccidia organisms flourish in warm, damp conditions, and can lead to an outbreak of coccidiosis. It is important to ensure that drinkers are not placed above a scratching area for this reason. Above a droppings pit is a good position, because any water spilled will then fall through the slats. Ammonia, easily detected by smell, can cause irritation of the lungs and nasal passages, leading to more serious conditions.

Large units will utilise an automatic scraper system to remove litter and manure, while small units will do this manually. It is a valuable commodity, with a potential use as an organic manure. Some companies arrange to remove it free of charge from large units, so that they can use it to compost, bag and sell to the horticultural trade. Small units, and those which only keep poultry for household use, frequently have kitchen gardens where the manure can be utilised. It is important, in this case, to compost it well before use, and not to let the poultry have access to it because of the danger of disease.

References

1 S. W. Gibson and P. Dun, *The Performance of Laying Fowls in a Covered Straw Yard System*, Technical Notes No. 162, West of Scotland Agricultural College, 1984–5.

3 LAND MANAGEMENT

'The poultry-man who adopts free-range methods should move the houses frequently and keep the grass short.'

— Leonard Robinson, 1948

The key to successful free-ranging is good land management. This applies to the nature of the land itself, the degree of shelter it offers, how it is fenced to deter predators and how pasture is managed and maintained.

THE NATURE OF THE LAND

It is no coincidence that the great free-range egg-producing areas of the past were in relatively mild areas blessed with free-draining soils. The Lancashire sands and Wiltshire chalks were ideal, providing land free from boggy areas, although in very hot summers there could sometimes be problems of grass scorching in particularly thin-soiled areas. The ideal is to have adequate drainage to prevent waterlogging, while ensuring that the bulk content and fertility of the soil are sufficient to retain and provide enough water and nutrients for a healthy growth of grass. The balance is not always easy to achieve, but it is better to start with a well-drained soil than a heavy one.

It is possible to improve drainage, but this can be an expensive option. Where the water table is naturally high, the only long-term solution may be to plough the whole area and excavate drainage ditches, as has been done in parts of the Fens. It is possible to install permanent tile drains, but this is extremely expensive and difficult to justify in terms of the likely returns. In waterlogged areas, where the water table is not normally high, the problem may be one of 'panning'. This occurs when the top few inches of a soil have become so compacted that the surface pan holds the water, without allowing it to drain through. It is usually clay soils which are affected in this way, although lighter soils may become panned if overuse has led to excessive compaction. The solution here is to mole plough the area. The 'mole' in question is a tractor attachment which breaks up the hard surface and forms a series of tunnels in the subsoil. There are agricultural contractors who will undertake this.

Once the area is ploughed, an application of lime in the autumn will help to flocculate the soil, the process in which tiny particles of clay clump together to make bigger particles, so providing larger air spaces for more efficient drainage. The addition of well-rotted manure later in the season will help to provide bulk and fertility, while a new grass ley mixture can then be sown to provide future pasture for the birds.

Where waterlogging is only apparent in small areas, the easiest solution is to dig a hole and place clinker at the bottom to make a soakaway. The point has been made elsewhere in this book that, where fixed houses are used, the areas immediately around the pop-holes should be concreted or provided with some kind of verandah to avoid muddy conditions from the birds' scratching activities.

SHELTERED AREAS

Land which is to support free-ranging chickens adequately must be sheltered. This does not refer to the obvious need for a house, but to factors such as the availability of trees, walls, windbreaks and hedges. The chicken originated in subtropical forests where trees provided cover and protection from wind, rain, sun and predators. It is out of its natural environment in a wide open field, and tends to stay near the house. Consequently, the grass immediately around the house soon deteriorates, while that further afield is barely touched. Although European Community regulations demand that the stocking density does not exceed 1,000 birds to the hectare, the actual density in the limited area the birds tend to frequent is usually far higher than this. It is a great problem to those who have invested in large, fixed houses, as has already been mentioned.

The West of Scotland Agricultural College has conducted a survey to investigate the distribution of land usage and the number of birds going out onto range. By plotting the location of birds at one-hourly intervals, they discovered that 55% stayed in the area

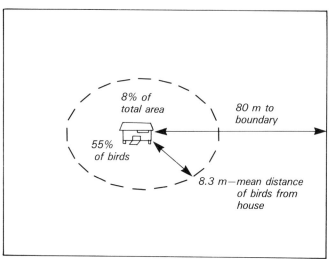

Drawing not to scale
Based on information from the West of Scotland Agricultural College

Figure 3.1   Distribution of land usage by free-ranging birds

Part of the author's free-range flock in an orchard where the tree cover provides shelter and a sense of security.

around the house, thereby using 8% of the total area available. The mean distance of birds from the house was 8.3 m out of a possible 80 m.[1]

Weather is obviously a factor which influences chickens in using the range, with warm, dry conditions being more conducive to ranging than wet, windy ones. However, the shelter/security factor is not one which appears to have been investigated to date. My own observations are that chickens will range much further afield if there are plenty of trees on the site. Keeping chickens in an orchard proved to be the ideal, with a far more uniform distribution over the whole site than when I had them in an open field.[2] This goes against the Ministry of Agriculture recommendation to cut down trees on the grounds that they provide fox cover, a claim which I would dispute. If a site is properly fenced against foxes, the provision of trees is an excellent inducement to the chickens to range over the whole protected site. Trees also provide much-needed shelter from wind, although any which are too near the perimeter fence may need to be cut back if they are likely to provide a launching pad for chickens to flutter over.

Trees are not the only form of shelter, of course. Wattle hurdles, hedges, banks, straw bales and stone walls all provide protection from the wind, as well as the feeling of security which is so necessary for chickens if they are to forage all over the available ground. It is interesting to note that a free-range unit which was set up in an old 3-acre walled garden experienced no problems in getting the chickens to range all over the site. There were several trees, and the high stone walls gave an overall feeling of enclosure.[3] This is in complete contrast with, for example, the National Agricultural Centre's unit which, on the date of my last visit there in May 1989, had a flock of over 400 birds in a fixed house with surrounding, treeless pasture. I counted 18 chickens outside, with only 3 moving away from the cluster around the wall of the house. All the others were inside at 2.35 pm on a warm, sunny day with no wind.

Moveable shelters are recommended for use with outside drinkers in hot weather. Such shelters need not be elaborate or expensive constructions, merely something that will provide shade from the hot sun and help to keep the water supply relatively cool.

FENCING

Traditional wisdom has always valued the adage, 'Good fences make good neighbours.' As far as the free-range enterprise is concerned, good fences also make good sense! Without them, the depredations of the fox would make a mockery of any effort to keep poultry. In addition, wandering dogs, feral cats, mink, badgers and even polecats in some areas, may prove hazardous to chickens.

Fencing is also required to control the birds' access to pasture, so it is appropriate to regard fencing in two ways: external or perimeter fencing to keep out predators, and internal or pasture control fencing to restrict the movement of chickens to certain areas.

Perimeter fencing

This should separate the poultry farm from the outside world. There is no easy or cheap way of doing this and it is one of the most substantial costs facing anyone thinking of starting a free-range poultry farm.

A 2 m perimeter fence will provide adequate protection, as long as there is a further overhang of 30 cm placed at an angle of 45° to the vertical. The overhang should project outwards to repel boarders, and the fence posts will need to be inside the fence itself so that no convenient footholds are provided. Wire mesh netting with 50 mm holes in the mesh will deter foxes, although mink can still climb a fence and wriggle through holes of this size. However, mink are more likely to be a problem in areas close to rivers. In such situations, trapping may need to be considered, and the Ministry of Agriculture should be consulted for advice on suitable traps and procedures. In some areas it is possible to borrow traps from the authorities.

The bottom of the wire mesh needs to be dug well into the ground to prevent predators from pushing their way underneath. The provision of an electrified wire will give added security to the fence, in this respect, if it is placed 25 cm high and 25 cm further out from the mesh. Placing a second wire near the top, or just above the fence, 10 cm out from the mesh, will make the barrier virtually invulnerable. The details are shown in Figure 3.2. A high power mains fencer unit is recommended for a permanent fence of this kind, because it needs less maintenance than a battery operated one. Wiring should

54

never be connected directly to the mains supply and the advice of a specialist should be sought before installing such a system. There are many suppliers of electric fencing who are experienced in the needs of poultry-keepers, and who supply fences which conform to the British Standards Safety Requirements.

At the time of writing, the cost of erecting an electrified fence of the type described above is approximately £2 per metre. If an existing fence is electrified, the cost is about half this sum per metre.

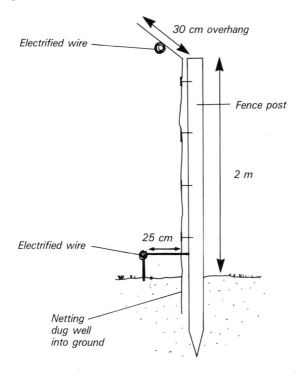

Figure 3.2 Perimeter fencing

Pasture control fencing

This is to control the access of birds to certain areas of pasture, rather than to keep out predators. It does not need to be particularly high – 90 cm is usually sufficient – but it should be easily moveable and re-erected as required. Electrified netting is ideal for this purpose, although it should be added that it is also suitable for incorporation in a permanent perimeter fence. It is essentially a series of lightweight plastic fencing posts with metal spikes which are tapped into the ground. These are non-conducting and are purely for support, with one being placed every 3 m or so. The netting is normally made up of 8 horizontal lines of heavy gauge polythene/stainless steel electroplastic twine, with non-conducting polythene verticals and bottom horizontal strand. The fence is tensioned with straining post guys and pegs, and powered from a mains operated unit or a battery operated one. The latter is normally a 12 volt rechargeable car battery. When the fence is moved, it is uprooted, rolled up and re-erected on the new site.

Electric netting used to control access to specific areas of pasture. (Bramley & Wellesley Ltd)

No system is without its problems, and the chief ones are shorting and failure of the power supply. Shorting can occur if the grass gets too long where the fence is positioned, so it is important to keep that area mown. A battery powered unit also needs regular checking in case it needs recharging. A neon tester or electric voltmeter for checking the state of electrification is highly recommended.

Another problem which may arise is tangling of the netting when it is moved. It is much easier for two people to move the fence and to ensure that it is evenly rolled.

56

PASTURE MANAGEMENT

A field is often regarded as a permanent and unchanging entity. In one sense this is correct, but if the grass is regarded as a crop, it is obvious that it is temporary and must be managed properly on a seasonal basis. Chickens cannot be allowed to range on the same piece of land indefinitely, otherwise the grass will deteriorate, there will be a gradual increase in the incidence of pests and parasites, and the flock will succumb to major health problems.

ADAS recommends that chickens should generally be moved to a new area of vegetation every 4–6 weeks, if a system of paddock grazing is used. With a moveable house this poses no great problem; it is simply a matter of transporting it by hand or tractor, depending on its size, and re-erecting the electric netting, pasture control fence around the new area. Where only a small number of birds is kept in a house with combined run, it may be more appropriate to move it every day or every few days. If the flock is ranging in numbers well below the official limit and has a considerable expanse of pasture at its disposal, it may not be necessary to move it as frequently. For example, a house may have a field on either side of it, with these two areas being used alternately. It is a question of relative scale, stock density and common-sense, but with a commercial unit, the ADAS recommendation is a good, general guide.

The area immediately around the house becomes over-used, and efforts must be made to encourage grazing further afield. (Shaver Poultry Breeding Farms)

57

Where fixed houses are used, the birds' access to new grazing will be controlled by closing off the pop-holes to the old pasture, and opening up those leading to the new one. Again , a portable electric netting fence is the ideal way of controlling the flock's ranging once it is outside.

After a flock has been moved to a new area, the old pasture should be raked if there are any patches of compacted droppings, and then 'topped' to cut down taller grasser which may be producing seed heads. An alternative is to graze sheep, cattle or goats on the site; they will do all the necessary topping. Another course of action is to let the grass grow anyway, after the birds have been removed, and then to cut the resulting growth as a hay crop. If other grazing livestock are kept, it can be a useful source of fodder. Some free-range units rent out such land to farming neighbours who require extra grazing, while others graze sheep at the same time as the birds. There are some who claim that this practice may lead to cross-infection, which adversely affects the birds, but there is little evidence of this, as yet. It is a point on which local veterinary advice should be sought.

On a large scale, topping can be done with the traditional farm equipment of tractor and cutting bar. A ride-on garden tractor with the blades set at maximum height is also effective. For small areas, an ordinary hand lawn mower with the blades set high will do the job.

Mention has already been made of the need to keep grass short near electric fencing. Regular mowing is the only way to ensure this and a ride-on or hand garden mower is more effective and less cumbersome at keeping a strip close to the fence clear of tall growth. Remember to switch off the electric fence before mowing around it!

It may also be necessary to mow the area of pasture currently occupied by the birds, if it is growing more quickly than their foraging actions can keep pace with. Tall grasses are not eaten by the birds and rank growth, particularly if wet with rain or dew, is a positive disincentive to them. Those which do brave such conditions get their legs and bottom feathers wet, making eggs which are subsequently laid wet and dirty.

Once birds have vacated a site, and if other livestock is not being grazed on it, it is a good idea to dust with lime, particularly if the ground tends to be on the heavy or acidic side. There is some dispute among poultry experts whether this practice helps to deter disease-causing organisms and clear up parasitic infection of the

A close watch should be kept on pasture to ensure that the grass does not get too long or that bare patches do not develop.

ground. Traditional wisdom has always encouraged it – and what is certainly true is that heavy, clay soils are less likely to compact and puddle if they are treated with lime. A reduction in the number of waterlogged areas will also reduce the incidence of snails which act as the intermediate hosts to parasitic flukes and the coccidiosis-causing organisms. Liming was a practice which I always followed to good effect with my own pasture rotation management.

If an area of pasture has deteriorated badly, it may be appropriate to plough it up, harrow it and then reseed it with a new ley mixture. Once it is ploughed, test the soil to determine its pH level of relative acidity or alkalinity and apply lime if necessary. Leave it to weather for a time, then harrow it to break down the clods of earth ready for seeding. All these activities can be carried out by a specialist contractor, if you are not in a position to do it yourself.

Special ley mixtures for free-ranging poultry are available from some specialist seed suppliers. These mixtures are made up of shorter, perennial grasses which are more suitable for poultry than the longer grasses which are usually found in other leys. As a general rule, 50 g per square metre, or 500 kg per hectare, of seed will be required.

A new ley pasture will not generally require feeding in its first year. In the second year it can be given some nitrogen fertiliser, and it is best to use one of the more environmentally acceptable ones such as calcified seaweed which will not leach into water courses.

References

1 Linda Keeling and Peter Dun, *Polythene Housing for Free-range Layers: Bird Performance and Behaviour*, Scottish Agricultural Colleges Research and Development Note No. 41, February 1988.
2 Author's unit at Widdington, Essex, 1975–88.
3 *Poultry World*, June 1987.

4 BREEDS

> 'In order to make a success of any branch of
> poultry-keeping, it is important to keep the right breeds.'
> — Herbert Howes, 1939

If the aim is to produce the maximum number of eggs possible under free-range management, the choice of bird will be one of the modern hybrid strains bred for egg production, but if poultry is to be raised for the table, this would be the last choice one would make. If the main interest is in breeding show birds and selling stock to other poultry-keepers, the old traditional and pure breeds would be the obvious choice. It is important to keep the right breed for the job and not to confuse the various issues.

BREEDS FOR EGG PRODUCTION

For free-range egg production, a modern hybrid layer of brown eggs is the best choice. British consumers prefer brown eggs, perhaps seeing them as more wholesome and associated with the farm. There is no truth in this belief! The factor which determines shell colour is a genetic one. If a hen which lays white eggs, such as the Leghorn, Wyandotte or Ancona, is kept on free-range, it will continue to produce white eggs, just as a brown layer will produce brown eggs in a battery cage. Most distributors, retailers and some free-range feed manufacturers who provide marketing help insist on the production of brown eggs if their posters and marketing aids are displayed.

The best choice for free-range egg production is one of the modern hybrid strains such as ISA Brown, Hisex Brown, Shaver Brown, Hy-Line Brown, Babcock Brown or Hubbard Golden Comet. These are all layers of brown eggs and have been selectively developed in the past from the old American breed, Rhode Island Red. A hybrid is the product of several different breeds, the progeny of which

61

ISA Brown hybrid layer. (ISA Poultry)

demonstrate particularly good characteristics such as number of eggs laid, size and quality of eggs, etc. and are selected for further breeding. A modern hybrid is the result of these various strains being brought together, and is highly bred for efficient production.

It is appropriate to ask why, if Rhode Island Red is the basis of all these brown egg hybrids, this breed is not the first choice for egg production now. It is certainly true that in egg laying trials in the past this breed was recorded as having produced over 300 eggs a year – but that was over half a century ago! Most of the pure breeds that are available now have been bred for show, not for their productive capabilities. Utility rather than show strains are available from some breeders, but they have been crossed with other breeds and strains to reintroduce hybrid vigour and productive qualities. They may be called Rhode Island Reds, if this is what they are based on, but strictly speaking they are not the old pure breeds, but updated versions of the original. The old Rhode Island Red has deep, mahogany brown feathers and is a much bigger bird than its modern counterpart, which is smaller and has lighter red feathers. It is usually only the show breeders who are now selling the real pure breeds; most other sources are selling hybridised crosses and either calling them by the original name or using a term such as 'modern Rhode Island Red'. The latter is, in effect, a hybrid! Some suppliers sell first crosses such as Rhode Island Red × Light Sussex or the popular Black Rock, which is a cross between the Rhode Island Red and the Barred Plymouth Rock.

A Rhode Island Red × Light Sussex layer, a popular traditional choice but not as productive as a modern hybrid.

Where the older, more traditional breeds such as Rhode Island Red, Sussex, Wyandottes, Leghorns and their various crosses are chosen, they should be 'utility' or productive strains rather than show birds, unless the latter are particularly desired. Even so, they will tend to produce fewer eggs and consume more food than the modern hybrids.

Modern hybrid layers have been largely bred for the battery industry, but experience over the last decade has shown them to be quite capable of adapting to free-range, although no chicken can cope with cold, wet and exposed weather conditions. Some people claim that the older breeds are hardier in outside conditions, but there appears to be no recorded evidence to support this. Recent findings do, however, indicate that pullets destined for free-range conditions need to be heavier than those going into cages, otherwise they may

Marans are a popular choice if dark brown, speckled eggs are required.
(Walner)

not be able to eat enough for body metabolism, energy requirements and good sized egg production. A body weight of 5–10% over that of caged birds is required for free-range, point of lay birds.[1] Natural, floor rearing conditions without giving extra light will ensure proper development without giving rise to precocious birds.

Where hybrids have not yet been able to compete with some of the old breeds is in the production of very dark brown, speckled eggs. A breed such as the Maran or the Welsummer will produce beautifully dark eggs, as long as the bird is a good example of the breed. Inferior specimens will lay eggs that are not much different from any other brown egg. Top class examples really are almost chocolate brown, with ample speckles, and can command a premium price. Some free-range units keep flocks of normal hybrid layers for quantity of eggs, and smaller flocks of Marans or Welsummers for 'top of the range' sales. Such eggs sell themselves!

Table 4.1 indicates some of the brown egg breeds which have been kept in the past, as well as the modern hybrids of today.

The choice of bird may be largely decided by which supplier is most conveniently placed for a particular area. There are obvious advantages to having the supplier on the door-step, and if any problems arise, they can be readily sorted out through direct contact. What

Table 4.1 Brown egg breeds

Hybrids	Pure breeds or first crosses
ISA Brown	Maran
Hisex Brown	Rhode Island Red
Shaver Brown	Barred Plymouth Rock
Hy-Line Brown	Black Rock
Babcock Brown	Rhode Island Red × Light Sussex
Ross Brown	Dorset
Hubbard Golden Comet	Welsummer

is important to establish is the production record and the egg weight profile. The former will be evidence of a bird's parentage and productivity, indicating the number of eggs it is likely to produce. The latter indicates the average number of size 1, 2 and 3 eggs it is likely to lay. These factors are crucial to a free-range producer, particularly the egg weight profile, because large eggs fetch the highest prices. The ideal free-range egg layer has the following characteristics:

- A good hen-housed average for the number of eggs laid.
- A high proportion of large eggs.
- A moderate feed intake in relation to its size.

Most of the modern brown egg hybrid layers will meet these requirements, and it is personal preference and circumstance which determines what is ultimately selected. Contact the various breeding companies and ask them to send details of egg production and egg weight profiles for their breeds. Table 4.2, for example, shows the results achieved with a flock of 3,000 ISA Brown hybrids kept on free-range, with the hen-housed averages quoted at 72, 76 and 100 weeks respectively. Table 4.3 shows the egg weight profile of the same flock.

Table 4.2 Production record of 3,000 free-range ISA Brown flock at Hillside Farm, Berkshire

Hen-housed average to 72 weeks	273.03
Hen-housed average to 76 weeks	295.35
Hen-housed average to 100 weeks	412.35
Feed conversion kg per dozen eggs	2.12
Liveability	92.7

Table 4.3   Egg weight profile for 3,000 free-range ISA Brown flock at Hillside Farm, Berkshire (weight in g)

Age in weeks	Size			
	1	*2*	*3*	*4*
40	7.2	19.8	36.8	27.3
50	15.1	31.8	38.4	9.7
60	21.6	34.4	31.3	7.2
70	34.9	35.2	23.9	3.2
80	30.8	39.3	22.9	4.4

Source: *ISA Brown Newsletter*, No. 5, May 1989.

BREEDS FOR THE TABLE

Traditionally, the Light Sussex was the favourite breed for table production, although many of the old heavy breeds such as the Dorking, Indian Game, Cornish, Barred Plymouth Rock and their various crosses have been used. A particularly popular cross at one time was the Light Sussex × Indian Game, using either a Light Sussex male on Indian Game females, or vice versa. The Indian Game has a particularly broad breast, a factor which is carried through to the crossed progeny.

Light Sussex cock and hens, one of the traditional table bird breeds.

66

Another favourite cross was the Rhode Island Red male with Light Sussex females. This gives a sex-linked cross, where males and females can be identified at birth; male chicks are silvery yellow, while females are brownish orange. The RIR × LS was regarded as a good dual-purpose bird, being suitable for egg production as well as for the table. Surplus males were usually reared for the table, while females were kept as replacements for the laying flock.

In recent decades, with the divergence of selective breeding into the production of layers or broilers, dual-purpose breeds have become commercially redundant, although they have always retained their popularity in the smaller sector. The emphasis has been on breeding white-feathered, quick-growing birds, of which the Cobb is the prime example. Most of the breeding companies which produce laying strains also breed their own white-feathered broilers such as Hubbard broiler, Shaver Starbro and Ross broiler. These hybrids have all been developed for intensive, indoor conditions, but they are quite capable of less intensive management. I regularly raised Cobbs for the table, keeping them in a poultry house with an attached straw yard, and allowing them to range on grass when weather conditions permitted. They grow more slowly in this way, but prove quite acceptable for household production.

The emphasis on developing white-feathered hybrids was mainly because, after plucking, there are no dark stubs left in the carcass, and it was perceived that this was what the consumer demanded. In recent years there has been an interesting reversal of this, with darker-feathered birds becoming associated with non-intensive production. In France such birds are called by the general name *Label Rouge* (Red Label), a reference to the trade description under which they are sold, as well as to the fact that the birds are red-feathered. With this description, consumers can identify them as being more naturally reared. One of the most common examples in France is the Loué.

These red-feathered birds are more slow growing and have been developed for non-intensive production, with most of their diet coming from grain. British breeding companies are now producing their own versions of these strains. Examples are the Shaver Redbro and the ISA 657 Red-feathered Broiler. If the aim is to specialise in producing table birds for the non-intensive market, it is well worth considering one of these breeds. The smaller producer could also consider one of the traditional crosses such as those referred to

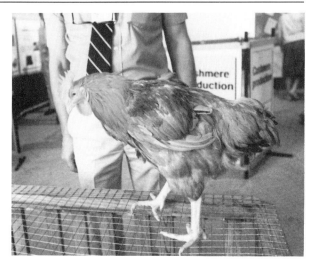

Shaver Redbro cockerel, a slow-growing broiler strain suitable for free-range production.

earlier. It would be satisfying to see flocks of Light Sussex × Indian Game grazing the stubble fields of arable farms again.

Table 4.4 lists breeds and crosses which are used commercially, or have been used in this way in the past.

Table 4.4   Breeds of table poultry

Modern hybrid breeds	*Old breeds and crosses*
(White-feathered)	(White-feathered)
Cobb	Light Sussex
Shaver Starbro	White Cornish
Marshall	(Red-feathered)
Arbor Acre broiler	Dorking
(Red-feathered)	Indian Game
Shaver Redbro	RIR × LS
ISA 657 Red-feathered broiler	LS × IG (or vice versa)
	Plymouth Rock

KEEPING OLD BREEDS AS A HOBBY

There is a wide variety of pure breeds in existence, although some are now so rare as to border on extinction. It is thanks to the dedication of small breeders that some breeds are in existence at all. Rare breeds may be considered an irrelevance by some people, but it should be remembered that they represent an invaluable genetic pool which one day may need to be drawn upon, quite apart from their intrinsic value as individual and unique species.

The Poultry Club of Great Britain is a national organisation which sets recognised standards for the different breeds. There are also affiliated breed clubs which look after the interests of individual breeds. If you are interested in a specific breed it is a good idea to join such a club and establish what the standards are before acquiring stock. In addition to large fowl, there are also bantams which are popular with many people.

Most of the traditional breeds of chickens can be seen at specialist farm parks where there is an extensive collection such as that of the Domestic Fowl Trust.

BREEDS FOR THE FAMILY UNIT

If the aim is just to have a few chickens in the garden, to supply eggs for the family, it does not matter a great deal which breed is chosen, beyond the fact that a broiler bird would not be a sensible choice if eggs are the aim. Choose which breed you happen to like. If it is one of the pure breeds, you will be helping to support the breed and keep it in existence. The old breeds will cost more to feed and they will lay fewer eggs, particularly over the winter period when they may stop laying completely. In view of this, you may prefer to opt for a modern hybrid. If space is really confined, consider keeping a few bantams.

None of the eggs can be sold unless you have the flock tested regularly for salmonella, and the costs of this are greater than could be covered from small-scale egg sales. You should not allow the number of birds you keep to exceed 24, otherwise the whole flock will need to be tested, regardless of whether you are selling eggs.

PURE BREEDS OF POULTRY

Pure breeds are those which will breed true. In other words, the progeny will resemble their parents and have the same general characteristics.

Pure breeds are generally differentiated into light, heavy or dual-purpose breeds. Light ones are light in weight and usually lay more eggs than the heavier birds. Heavy breeds are more suitable for the table, or as 'sitters' for hatching eggs. Dual-purpose breeds can be used for both purposes, and the classic example is the Rhode Island Red. The following breeds are those which have been raised as domestic poultry in various parts of the world but, these days, are kept predominantly as show birds rather than for their productive qualities. Within many of the breeds there is a considerable variety of plumage colouring. For example, the Wyandotte has the following varieties: Barred, Black, Blue and Blue Laced, Buff and

Silkie hen, a popular choice of the fancy fowl breeder.

Table 4.5 Pure breeds of poultry

Breed	Type	Origin	Eggs
Ancona	Light	Italy	Creamy white
Andalusian	Dual	Spain	White
Appenzeller*	Light	France	White
Araucana	Light	Chile	Blue green
Aseel	Heavy	India	White
Australorp	Heavy	Australia	Brown
Barnevelder	Heavy	Netherlands	Brown
Brahma	Heavy	India	Brown
Bresse*	Heavy	France	White
Campine	Light	Belgium	White
Cochin	Heavy	China	Deep brown
Creve-Coeur*	Heavy	France	White
Dorking	Heavy	UK	White
Faverolle	Heavy	France	Cream
Hamburgh	Light	UK/Netherlands	White
Houdan	Light	France	White
Indian Game	Heavy	UK	Brown
Ixworth	Heavy	UK	White
Jersey Giant	Heavy	UK	Tinted
Lakenfelder	Light	Holland/Germany	White
Langshan (Croad and Modern)	Dual	China	Dark brown
Leghorn	Light	Italy	White
La Fleche	Heavy	France	White
Malay	Heavy	Asia	Tinted
Maline	Dual	Belgium	Pale brown
Maran	Dual	France	Dark, speckled brown
Marsh Daisy	Dual	UK	Tinted
Minorca	Light	Spain	White
Modern Game	Light	UK	Tinted
Naked Neck	Light	Hungary	White
New Hampshire Red	Heavy	USA	Tinted
Norfolk Grey*	Dual	UK	Tinted
North Holland Blue	Dual	Netherlands	Brown
Old English Pheasant Fowl	Heavy	UK	Tinted
Old English Game	Heavy	UK	Tinted
Orloff	Heavy	Russia	Tinted
Orpington	Heavy	UK	Brown
Plymouth Rock	Dual	USA	Tinted
Poland	Light	Poland	White
Redcap	Light	UK	White
Rhode Island Red	Dual	USA	Deep brown
Scots Grey	Light	UK	White
Scots Dumpy	Light/Sitting	UK	Tinted
Sicilian Buttercup	Light	Italy	White
Silkie	Light/Sitting	China	White
Spanish	Light	Spain	White
Sultan	Heavy	Asiatic	White
Sumatra Game	Heavy	India	White
Sussex	Heavy	UK	Tinted
Welsummer	Light	Netherlands	Brown, speckled
Wyandotte	Dual	USA	Brown
Yokohama	Light	Japan	White

* Rare breeds.

71

Buff Laced, Columbian, Gold Laced, Partridge, and Red and White.

Anyone considering keeping old breeds is advised to contact the secretary of the appropriate breed club who will advise on the various standards for the breed, as well as indicating breeders who have stock for sale. (See Reference section.)

In confined areas bantams may be more appropriate.

BANTAM BREEDS

True bantams are naturally occurring small birds which have no counterparts among the larger breeds. Developed bantams (my term) are those which are, strictly speaking, not true bantams, but which are now usually found only in bantam form or which have been recognised as such for a long time. Miniature fowl are large fowl which have been selectively bred in order to produce small, scaled-down versions of the original.

Where space is limited, bantams are ideal, although they will tend to lay eggs in spring and summer only, unless the winter is particularly mild.

72

Table 4.6 Bantam breeds

True bantams	Developed bantams	Miniature fowl*
Nankin	Booted	Ancona
Rosecomb	Frizzle	Araucana
Rumpless	Pekin	Australorp
Sebright	Japanese	Leghorn
Tuzo	Belgian Bearded	Maran
	Old English Game	Light Sussex
		Rhode Island Red

* Most of the large breeds now have scaled-down versions. This lists the most popular types.

AGE AT WHICH TO BUY STOCK

The best time to buy laying stock is at the age of 18 weeks when they are at the 'point of lay' (POL) period. Laying normally starts from 20 weeks and continues for between 52 and 60 weeks until the period of moult. Acquiring the young birds a couple of weeks before commencement of lay gives them time to settle down and to become accustomed to their new home.

A 17-week-old Hisex Brown pullet just before being moved from the rearing to the laying house. (Mick Dennett. Bibby Agriculture)

It is possible to breed your own replacement stock, but this is quite a different enterprise from that of egg production. A breeding flock must be kept quite separate from the laying flocks, and all breeding birds should be tested to ensure that they are healthy, vigorous and free from genetic defects and transmittable diseases. Any flock of over 24 birds must be registered as a breeding flock and regular testing for salmonella is required, as mentioned earlier. (See also Chapter 10.)

It may be appropriate to buy broilers as day-olds when they are at their cheapest, as long as heated conditions are available for them. If they are bought at around 6 weeks old, they will be hardier, will not require heated conditions and the mortality rate will be lower, although they will be more expensive than day-olds. As a general rule, they will be reared for a minimum of 13 weeks before slaughtering.

TREATMENT OF BOUGHT-IN STOCK

Any stock which is bought in should have been non-cage reared and fully vaccinated against Marek's disease, Newcastle disease and infectious bronchitis. They should also have come from breeding stock which has been tested and found to be free of salmonella. Check with your veterinary surgeon or the local Ministry of Agriculture as to what precautions may be necessary in your particular area in case it is necessary to vaccinate against conditions such as endemic tremors as well. Once you have bought the birds, you will also be required to test the birds for salmonella on a regular basis if eggs are being sold, or if you have a large breeding flock.

The quarters should have been prepared in good time before the arrival of the birds, with adequate supplies of food and water available. If a previous batch of birds has inhabited the house, it should be thoroughly cleaned and disinfected before use. Ensure that awkward corners and the edges of perches have been adequately treated, particularly if there has been an outbreak of lice or mite infestation.

Keep the new arrivals locked up for at least 24 hours before letting them out, so that they know where 'home' is. Merely turning them out onto pasture without introducing them to their house can cause serious disorientation and stress.

If the new arrivals are day-olds they will need carefully controlled environmental conditions, with artificial heat. Further details are given in Chapter 9 on Breeding and Rearing.

References

1 Shaver Poultry Breeding Farms, Technical Information Sheet, 1987.

5 FEEDING

'If the last portion of food necessary to satisfy a bird's appetite is missing, it is egg production which is likely to suffer.'

– Keith Wilson, 1949

The metabolism of the hen – like that of any other creature – includes a wide variety of activities. She grows, repairs her tissues, grows healthy new feathers, breathes, moves about, lays eggs, clucks, pecks and scratches. The list is almost endless, but these activities would soon be curtailed if food was not available to power them.

The basic requirements, apart from water, are proteins, carbohydrates, minerals and vitamins. All these foods cater for the overall metabolism of the bird but, as a generalisation, proteins may be said

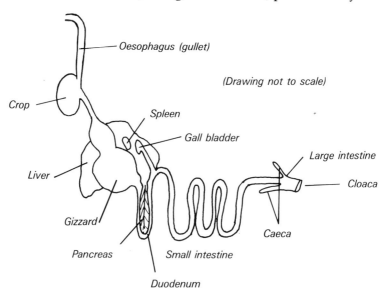

Figure 5.1 Digestive system of the chicken

76

to cater primarily for growth, carbohydrates and fats for energy, and minerals and vitamins for health. This is an oversimplification, but as a working definition for preparing a hen's feed ration, it is an adequate one.

PROTEINS

Proteins are complex substances found in both animal and vegetable sources. The main animal protein sources are skim milk powder, fishmeal, and meat, blood and bonemeal. The main vegetable sources are soya and other beans, maize and other cereals, and sunflowers.

Proteins are made up of constituents called amino acids. There are about a dozen of these but the most important are lysine, methionine and tryptophan. A hen is capable of synthesising most amino acids from other food constituents, but these three must be taken in directly every day. The average laying bird will require a daily intake of 900 mg lysine, 430 mg methionine and 200 mg tryptophan. The overall protein requirements are 18–19 g per bird, per day. Most compound feeds declare the protein content on the sack, and this is normally between 16 and 18%. A 16% ration would require a higher daily intake than the 18% ration. Compound feeds are normally fed on an *ad lib* basis, where birds can help themselves from feeders placed in the poultry house, but the intake needs to be balanced with a grain ration. This not only meets energy requirements but also keeps relative costs at an acceptable level. Compound feeds are obviously far more expensive than cereals.

ENERGY FOODS

Carbohydrates and fats are the important energy producers and these are found primarily in cereals and beans, although animal and vegetable protein sources also provide them. Energy in feedstuffs is referred to as its metabolisable energy (ME), and this is measured as mega joules per kilogram (MJ/kg). A laying bird will normally require 11.5 MJ/kg, but temperature has an important bearing on this.

The optimum temperature for a layer is 21°C, a reminder of its warm climate origins. In winter a free-ranging bird is asked to produce eggs in temperatures which can go down to below freezing point. For every 1° fall in temperature from the optimum, a laying bird will need an extra 4.2 calories. This necessitates a considerable increase in energy-producing feedstuffs and the feeding of extra cereals as a scratch feed in winter. Ensuring that a house is well insulated will go some way towards keeping the layers warm, but there is no alternative to extra winter feeding. Extra compound feeds will also make up the deficit, but the cost of this compared with that of grain makes it financially unviable, bearing in mind that the cost of cereals is far lower than that of compound feeds.

Figure 5.2 shows how increasing the MJ/kg energy value in the diet by feeding higher levels of grain prevents the wastage of protein which results from higher intake of compound feeds.

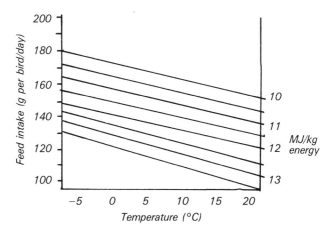

Source: Tony Warner, ADAS Reading, ATB Course, 1987.

Figure 5.2   Feed intake of free-range birds in relation to temperature

MINERALS AND VITAMINS

Minerals

Chickens require the normal range of minerals that most living organisms use, but the crucial ones are calcium and phosphorus, which play an important part in bone structure and eggshell quality. Compound feeds contain both these minerals, with around 4%

calcium and about 0.3% phosphorus. If such feeds are given, it is not necessary to add extra limestone in the form of crushed oystershell. This was recommended practice in the past, but mainly because accurately compounded feeds were not available and there was a common reliance on feeding scraps whose nutrient levels could not be assessed. On a small domestic scale, if a compound layer's feed is not used, it is still a good idea to make a little crushed oystershell available in the run.

Salt is an important part of the diet but should not exceed a level of 0.4%. Too little will adversely affect growth and egg production, while too much will cause excessive drinking, leading to digestive problems and loose droppings. Again, compound feeds contain the optimum level, normally 0.38%.

Trace elements are small quantities of essential minerals and include zinc, iron, copper, selenium, iodine and molybdenum. They are normally included in compound feeds.

Vitamins

Vitamins are organic compounds, as distinct from the inorganic minerals. Many are available via the grass-grazing activities of the chickens, while vitamin D is produced through the action of sunlight. Small quantities of essential vitamin supplements are included in compound feeds.

COMPOUND FEEDS

Compound feeds are rations which combine all the basic nutrients that are necessary in the bird's diet. Based on cereals, they contain mineral, vitamin and trace element supplements to provide a balanced formulation for dietary needs. Often referred to as 'mash', the feed is available in powder, crumb or pellet form. Most feed suppliers now supply compound feeds specifically for free-range flocks and these are given in association with a separate grain ration.

Consumer reaction to what was perceived as an unacceptably high level of additives in poultry feeds, together with the general growth in the free-range market, has resulted in the production of compound feeds with more natural ingredients. It is now possible to obtain feeds without antibiotic growth promoters, and with more natural ingredients such as marigold and red pepper extracts for enhancing the colour of egg yolks. A typical example of a compound feed produced for free-range layers is given in Table 5.1.

Table 5.1 Compound feed ration for free-range layers

	%
Protein	18.00
Oil	3.50
Fibre	3.50
Ash	13.00
MJ/kg (energy)	11.60
Calcium	3.60
Sodium (minimum)	0.12
Salt (maximum)	0.38
Lysine	0.88
Methionine	0.40
Linoleic acid	1.40

Compound feeds are fed in the house, either in manually filled tube feeders or in an automated chain system in a large enterprise. The dry layer's mash is made available on an *ad lib* basis, although it is wise to keep the actual level fairly low to prevent wastage. As a general guide, each bird will consume 130 g of layer's meal a day, although this will obviously vary depending on circumstances.

Compound feeds are also available for chicks, growers and table birds. These are referred to as 'starter rations', 'grower's feed' and 'broiler rations' respectively. They have different formulations to cater for specific needs and, again, are available with a more natural range of ingredients. The compound feeds which are given in association with grain are often referred to as 'balancer' rations because they provide the remaining nutrients to achieve an overall balanced diet. Following consumer reaction against the practice of using animal protein in poultry feeds, many feed suppliers now produce free-range compound rations based on vegetable proteins.

Feeders and drinkers should be regularly cleaned and disinfected to ensure that no pockets of possible infection, such as salmonella bacteria, are harboured.

GRAIN

Free-range birds normally consume up to 15 g of grain per bird per day. This provides the correct balance of nutrients with the compound feed ration. In winter this can be increased to 20 g per bird, per day, in order to compensate for the extra energy needed to keep warm in cold weather. The point has been made earlier that feeding extra grain is a cheaper alternative to making more layer's mash available. Barley is the cheapest grain. Other suitable grains are wheat or kibbled (chopped) maize.

The grain ration should be fed outside, as an incentive to wider ranging, in a different area every day. It can be deposited on the ground as a scratch feed, or given in grain feeders. One of the disadvantages of ground feeding is that wild birds and rodents may be eating at your expense, as well as introducing a health hazard. To combat this, purpose-made feeders which can only be accessed by chickens are available. One type has been mentioned earlier, with a pedestal which only the weight of the chicken can displace to make the grain available. Wild birds and rodents, with their lighter weight, are prevented from gaining access.

An outdoor feeder which will allow access to poultry but not wild birds or vermin. (Biodesign Products)

Table birds will require 70% of their diet to be grain if they are to meet the requirements of nationally recognised organic standards. The slow growing strains of red-feathered broilers have been developed for such a diet. Birds of this type are often referred to as 'corn-fed' and can be distinguished in the supermarkets by their yellow skins. This is a result of a maize-based diet. (Corn – as in sweetcorn – is the American word for maize.)

GRIT

Most free-ranging birds will pick up small stones and pieces of grit. These are essential for the efficient working of the gizzard, which breaks down grain. It is recommended that insoluble grit should be made available, either in small hoppers or on the ground, at the rate of 20 g per bird, per month. This will ensure that reluctant rangers or soils lacking stones will be taken into account. Insoluble grit for chickens is widely available from poultry suppliers.

Outdoor taps and water pipes can be protected against freezing by the use of a thermostatically controlled warming cable. (Robert Frazer & Sons Ltd)

WATER

Chickens, like all living creatures, need fresh, clean water available round the clock. This is not surprising when you consider that nearly 70% of a hen's weight is water, and an egg contains around 65% water. Twenty-five birds will drink about 5 litres of water a day in normal conditions, with consumption increasing in hot weather. A water shortage which continues for up to 5 hours will cause the birds to eat less and the eggs will be smaller as a result.

Large units will have an automatic water supply consisting of a header tank, feeder pipes and drinkers. The latter are usually suspended over the droppings pit area – never over the litter area, otherwise a breeding ground for coccidia is provided! Small units generally rely on manually filled drinkers.

As long as the header tank, pipes and drinkers are kept thoroughly clean, problems rarely materialise in summer. Winter can be a different story. Frozen pipes are a real menace and every effort should be made to lag water pipes effectively. One of the heating

Outside water supplies should be in the shade. Although the drinker shown is suspended above the ground to keep it clear of litter, the grass is too long for effective free ranging.

83

tapes shown in the plate below is an effective way of guarding against frost ravages. It is self-regulating in that it only heats up when the temperature approaches freezing point. It can be wound round the pipes and tap and, once switched on, can be left to come into action if the temperature falls drastically.

In really hot summers there may be a problem with outside drinkers unless they are shaded. The birds are unlikely to use them shown is suspended above the ground to keep it clear of litter, the grass is too long for effective free ranging.
if the water is hot and will anyway appreciate a shaded area for their own use. It is well worth putting up some kind of temporary and easily moved shading where the drinker is situated. One poultry farm I visited had a large beach umbrella placed for the benefit of the birds. It looked slightly incongruous but the important thing was that it worked!

KITCHEN SCRAPS

I am frequently asked the question, 'Should I feed my chickens on kitchen scraps?' If the eggs are being sold, such scraps should be avoided, not only because the nutritional value is usually low, but because they may pose a health hazard. Where just a few birds are being kept for the household, it is common-sense to give leftovers to the birds, as long as they are fresh and it is not done to excess.

6 POULTRY MANAGEMENT

'Sound poultry husbandry should include due
consideration of their behaviour.'

— A. H. Sykes, 1971

POULTRY BEHAVIOUR

One does not have to be an animal behaviourist to establish the sort of characteristics the average chicken demonstrates. Anyone with a modicum of common-sense, keen observation and a 'stockman's eye' understands that the chicken is a pecking, scratching, perching creature, subject to the 'pecking order' of its peer group. It likes to flap its wings and take dust baths. It is also descended from the jungle fowl of the Asian subtropical forests, and does not like cold, wet or exposed conditions.

Armed with this knowledge, one is in a good position to cater for the needs of chickens, without going to the extremes of the intensive battery brigade on the one hand, and some animal liberationists who would turn everything out to fend for itself, on the other.

Pecking

The chicken's beak is adapted for picking up seeds of grain or snipping the growing tips of young grasses and shoots. It is obviously important, therefore, to ensure that the food made available to it is of the appropriate size and condition. Pellets or mash and the normal grains are suitable, but sunflower seeds or maize grains may need to be kibbled (chopped) before feeding. Similarly, fresh, young pasture will be appreciated, while old, rank growth will not.

Scratching

The chicken's powerful feet with their long toes and sharp claws are ideal for scratching, as anyone who has attempted to keep them in

85

a garden will have discovered to their cost. Free-ranging lets them scratch as much as they feel inclined to do, but it is important to rotate access to land in order to prevent excessive damage to the grass sward.

Perching

Like those of most birds, the chicken's feet and legs are adapted for perching above ground. Any house should be equipped with perches, and new birds should be purchased from breeders who raise ground reared stock, as distinct from cage reared. The former will adapt to their normal conditions without any problems, but the latter may need to be taught how to perch, such is the perverting influence on natural behaviour of extreme, intensive practices.

Taking dust baths

Any bird takes dust baths in order to rid itself of external parasites. Free-ranging chickens will normally find a dry, sunny spot where a natural dust bowl of fine soil is formed. Where space is restricted, it may be necessary to make an artificial one, using fine sand in a shallow container.

If an outbreak of mites or lice does affect a flock, it is not sensible to rely completely on a dust bath. The individual birds, the houses, perches and the dust baths themselves should be treated with an appropriate dusting powder.

The pecking order

The pecking order is a hierarchy in which some birds are more dominant than others. Cocks, old hens or even the poultry-keeper may be regarded as the dominant force. Anyone who has gone into a poultry house or area where there is a flock of female chickens, particularly modern hybrids, will have seen the tendency that many have of squatting in a submissive way, or even allowing themselves to be picked up without fuss. This can be a positive advantage where removal of birds is necessary.

The negative side of the pecking order is that some birds may bully others. It is important to watch out for this, particularly where docile birds are prevented from feeding adequately at the troughs, otherwise egg production, for example, will decline. There should be enough feeders to ensure that crowding does not take place, and pop-holes should be sufficiently wide and numerous to cater for stress-free exit and access.

Feather- and vent-pecking, and even cannibalism, are extreme examples of the pecking order in action. The best approach is to avoid such problems by providing enough room for the birds, as well as adequate feed, water and grit. Boredom can also trigger bullying. Providing suspended greens in a house, run or paddock is one way of ensuring that there is something more interesting to peck at than some meek little bird.

There are those who debeak their birds – in other words, they remove the tip of the upper mandible so that damage to other birds is avoided. I am not in favour of such a practice, and never found it necessary with my flocks. When I came across the occasional bully, my practice was to separate the bird for a few days of solitary confinement. This normally has the effect of disorientating the culprit to the extent that the original pattern of behaviour is forgotten. If that did not work, I would cull the bird for the pot!

MANAGEMENT OF LAYERS

There are two aspects of poultry management – the daily routine and periodic tasks.

The first daily task is to open the pop-holes in the morning so that the chickens can come out. This is a good time to check the flock, looking out for any suspicious signs such as feather-pecking or other manifestations of bullying.

Feeders and drinkers will need to be attended to on a daily basis. They should be clean and well filled, although not to the brim or the birds will waste a certain proportion of it. Drinkers should be placed above the droppings area so that any spilled water does not fall where the birds can scratch. This will help to avoid damp areas where coccidia organisms can multiply. Access to water is important and an outside source needs to be provided while the birds are foraging. Newly housed birds may need to be shown where the water supplies are based.

Feeders and drinkers will also need to be cleaned periodically, to reduce the possibility of infection in the flock. How often this is necessary will be decided from experience, but a daily check is vital to ensure that all is well. For cleaning, hot water, detergent and a small scrubbing brush are suitable for hand-filled feeders and drinkers. Pipeline and nipple drinkers will need a different approach, with a suitable poultry disinfectant being used. Follow the recommendations of the equipment manufacturer.

Examine the nest boxes daily for any eggs and remove them. At the same time, check for the presence of any ground-laid eggs and remove any litter which looks as if it has been designated a nest. Nest box liners should also be checked and, if necessary, replaced. Any droppings there should be regarded as a source of possible contamination and removed immediately.

Adjust the ventilation in the house by opening vents or windows as necessary, and check the timer unit controlling the lighting. This is also a good time to look for any signs, inside and out, which may indicate the presence of rats. Long, black droppings and gnawed areas of wood almost certainly mean that rats have been trying to gain access. Immediate action should be taken to control them, following the information given on page 97.

Check the perimeter fence to make sure that there have been no breaches during the night, and use a neon tester or voltmeter to assess the current in the electric fencing. If the grass is getting high it should be topped, particularly in the area by the electric netting, otherwise shorting of the current may result. A visual check of the pasture where the birds are ranging should also be carried out daily, to determine whether it is necessary to move them onto fresh ground. If, for example, bare or muddy patches are appearing, the next area should be made available so that the first can be rested.

During the day, while the birds are ranging, eggs should be collected regularly. The more frequently this is carried out, the sooner the eggs can be removed to cool conditions. An occasional check should be made on the flock to make sure that there are no problems such as feather-pecking. The outside water supply should also be checked and given shade protection if it is particularly sunny.

Mid-afternoon is a good time to give the birds their grain ration, although the precise timing will depend upon individual management plans. It can either be put in grain feeders placed well away from the house to ensure wider ranging, or placed on the ground in a wide arc for the same reason. It is a good idea to vary the place every day, so that undue scratching is not confined to one area.

Once the chickens have returned to the house in the evening, the pop-holes can be shut and the house secured for the night. It is not usually a problem to get the birds back, although one or two

Hens basking in a straw yard between their house and the pasture area. Care must be exercised to ensure that they do not start laying there.

individuals may prove difficult, particularly if there are trees on the site. In summer, when the weather is warm, some birds may want to revert to their ancestry and perch on the branches instead of going into the house. If the perimeter fence is secure against foxes there is usually no harm in this, although it may encourage a tendency to lay eggs outside.

COPING WITH PROBLEM BIRDS

No matter how good the standard of overall management, there will always be the occasional problem. It is easy to forget that nothing in life is static, and that any system is subject to a wide range of fluctuating influences such as the weather, mite attack or changes in the diet. Chickens may be part of a flock but they are also individuals operating within a pecking order of behavioural tendencies. A good stockman's eye will often detect incipient problems before they manifest themselves.

89

Some of the main problems and how to cope with them are detailed in Table 6.1.

MANAGEMENT OF TABLE BIRDS

The management of table birds is generally the same as that of layers, although there will obviously be slight differences, as in the lighting regime. Many of the problems that are likely to be encountered are also similar, apart from obvious ones such as egg binding and prolapse difficulties which are more likely in layers. Information which is specifically geared to table birds is in Chapter 8 on Table Poultry, while disease prevention and control are discussed in Chapter 11.

LIGHTING

Day length has an important bearing on the egg laying cycle. Once the day begins to get shorter, the number of eggs is gradually reduced until laying may cease altogether. In the wild, eggs are laid in several batches during the increasing light periods of spring to summer, with the reproductive cycle being confined to that period.

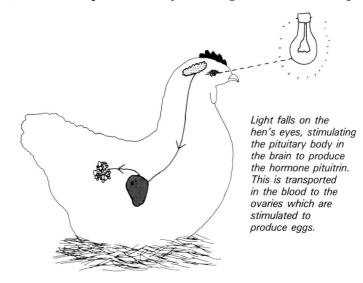

Light falls on the hen's eyes, stimulating the pituitary body in the brain to produce the hormone pituitrin. This is transported in the blood to the ovaries which are stimulated to produce eggs.

Figure 6.1 Effect of light on egg laying mechanism

Table 6.1 Problem birds

Problem	Cause	Remedy
Feather-pecking	Boredom.	Suspend cabbage greens for birds to peck.
	Inadequate diet.	Feed properly with balanced rations.
	Lice/mite attack.	Treat birds, house, nest boxes and perches.
	Overcrowding and stress.	Reduce stock density and review management practice.
	Light intensity too great.	Reduce light intensity by using dimming facility.
Vent-pecking	As above.	As above.
	Prolapse of the oviduct which attracts pecking by other birds.	More frequent in old hens where large eggs are laid. Avoid keeping old birds.
Cannibalism	As for feather- and vent-pecking.	As above.
Flighty birds	More likely to occur with some of the old breeds.	Keep modern laying hybrids bred for docility. Trim flight feathers on one wing in extreme cases.
Broody birds	More common in the old heavy breeds and in hybrids in their second year.	Keep modern layers and avoid keeping old birds.
	Partly an inherited characteristic.	Don't leave eggs lying around as an invitation.
	Likely to be worse in warm weather.	Clear sites likely to be used as unofficial nests.
		Remove culprit to cool conditions in a broody coop until she recovers, or to a hatching area if she is required to sit on a clutch of fertile eggs.
Egg eaters	See references in Table 7.1.	Collect eggs frequently.
		Use rollaway nest boxes.
		Try a mustard-filled egg to teach the culprit a lesson. Get rid of the really hardened criminals!
Feather loss	General loss.	Normal moulting occurs between egg-laying cycles.
	Excessive and prolonged loss.	Suspect depluming mite and treat with appropriate insecticide from vet. Ensure feeding is adequate.
Excessive feed consumption	Too cold.	Insulate house adequately. If necessary, increase grain ration.
	Worm infestation.	Treat with vermifuge from the vet. Change litter and rotate pasture more often.
	Feed being taken by vermin.	Check for presence of rats and mice and take control measures.

Although the domestic hen has been bred to produce eggs over a much longer period than her wild ancestors, she requires extra light if eggs are to be guaranteed through the winter.

The provision of light needs to be seen in two ways – day length and light intensity. Both these aspects play an important role and it is necessary to be able to distinguish between them. Day length is, quite simply, the number of hours in which light is available. The longest day is 21 June, when the natural day length is 17 hours. As this maximum declines, from July onwards, artificial light must be made available so that birds which are in lay do not have their day length shortened. Light intensity may be regarded as the strength or degree of brightness of light available. The unit of measurement is a lux, and light intensity is measured with a light meter. A satisfactory level of light intensity for a poultry house would be about 10 lux.

Lighting can be provided in one of several ways, depending upon the size of the house; 40 watt tungsten bulbs or 6–8 watt fluorescent bulbs or tubes are satisfactory, with the light sources placed 10 m apart. One light source is sufficient for up to 100 birds. On a small scale, an ordinary 25 watt bulb is satisfactory, while a portable system based on a 12 volt car bulb and battery suffices where mains electricity is not available. Whatever system is used it is important to incorporate a time switch so that the amount of artificial light made available can be controlled. A dimming facility is equally important so that birds are warned that the lights are soon to be extinguished, giving them time to find their way to the perches. It is also necessary to have a dimming facility to control the light intensity. If this is too great, it may, for example, encourage overactivity and stress, leading to cannibalism and other vices.

There are two golden rules with lighting. Do not provide light too early, before point of lay pullets have grown adequately, otherwise they will lay small eggs. The second rule is not to allow the day length to shorten once the birds are laying. Free-range birds must be adequately grown before they can cope with the demands of outdoor ranging as well as egg laying. If growers are approaching the point of lay in spring, there is usually no need to give them artificial light at all, unless they are housed in a particularly large house where natural light does not penetrate to a sensible degree. If light is required, it should be from the age of 20 weeks onwards, not 18 weeks as is often recommended, otherwise the bodily frame is not necessarily well developed enough for egg production and small eggs may result. Pullets which are housed from autumn to

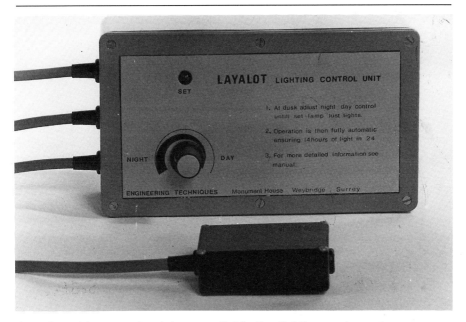

An adjustable control unit is essential in the lighting system. (Southern Pullet Rearers)

early spring should be reared on a constant 8-hour day length, which is equivalent to natural daylight at that period of the year. Pullets which are housed between spring and late summer need a stepped-down lighting pattern to ensure that they do not come into lay too early for their age. Recent trials established that a step-down lighting pattern starting at 23 hours at 1 week of age, reducing by 3 hours a week to a constant 8 hours, made the birds come into lay a week later than normal, but that there was a higher proportion of size 1 and 2 eggs between 24 and 32 weeks in lay.[1]

Pullets housed between November and February will need an increase of 15–20 minutes per week until the maximum of 17 hours a day is reached. Those housed between March and October, if they have had a stepped-down lighting pattern, can have an immediate increase of 2 hours, as long as they are no younger than 20 weeks. This is then followed by 20–25 minutes increase a week, up to the maximum. The lighting should be checked every day and adjustments made as necessary. It is appropriate to point out that many poultry-keepers prefer to follow the advice of animal welfare organisations, and do not exceed a maximum of 16 hours of light a day, outside the midsummer period.

Table 6.2 will assist producers to programme their lighting patterns according to the available amount of daylight hours.

Table 6.2   Daylight hours (Greenwich Mean Time)

Date	Sunrise	Sunset	Natural daylight hours	Date	Sunrise	Sunset	Natural daylight hours
Jan.				Jul.			
7	08.05	16.09	08.04	7	03.53	20.18	16.25
14	08.01	16.18	08.17	14	04.01	20.11	16.10
21	07.55	16.29	08.34	21	04.09	20.04	15.55
28	07.46	16.42	08.56	28	04.19	19.54	15.35
Feb.				Aug.			
4	07.36	16.54	09.18	4	04.29	19.43	15.14
11	07.24	17.07	09.43	11	04.40	19.30	14.50
18	07.10	17.20	10.10	18	04.51	19.17	14.26
25	06.56	17.32	10.36	25	05.02	19.02	14.00
Mar.				Sep.			
3	06.41	17.45	11.04	1	05.13	18.47	13.34
10	06.26	17.57	11.31	8	05.24	18.31	13.07
17	06.10	18.09	11.59	15	05.36	18.15	12.39
24	05.54	18.21	12.27	22	05.47	17.59	12.12
31	05.38	18.33	12.55	29	05.58	17.43	11.45
Apr.				Oct.			
7	05.23	18.45	13.22	6	06.09	17.27	11.18
14	05.07	18.56	13.49*	13	06.21	17.11	10.50
21	04.52	19.08	14.16	20	06.33	16.57	10.24
28	04.38	19.19	14.41	27	06.46	16.43	09.57
May				Nov.			
5	04.25	19.31	15.06	3	06.58	16.30	09.32
12	04.13	19.42	15.29	10	07.10	16.19	09.09
19	04.04	19.52	15.48	17	07.22	16.08	08.46
26	03.55	20.01	16.06	24	07.34	16.01	08.27
Jun.				Dec.			
2	03.49	20.09	16.20	1	07.46	15.55	08.08
9	03.44	20.16	16.32	8	07.55	15.52	07.57
16	03.44	20.19	16.35*	15	08.02	15.52	07.50
23	03.44	20.22	16.38	22	08.07	15.54	07.47
30	03.48	20.20	16.32	29	08.08	15.59	07.51

* Any free-range flock over 34 weeks of age by 13 April should receive natural progression of light to midsummer day (21 June). After 21 June they should be receiving 16 hours light until depletion of flock.
Source: J. Bibby Agriculture Ltd.

94

MOULTING

Moulting, where old feathers are lost and replaced by new ones, is a natural process which, in the wild, takes place once a year, in the summer. In domesticated chickens autumn- and winter-hatched birds will moult between July and August. Those hatched after March will normally continue through to October or November before moulting starts. It usually lasts a few weeks and egg production declines, and may even cease, while it is going on. The laying hybrid which has been bred for maximum production is less likely to cease laying all together than the older, pure bred bird. With several flocks of different ages the moulting periods occur at different times so that any decline in the number of eggs is minimised when the overall number of birds is considered.

It used to be the custom to 'force moult' laying flocks so that they would have moulted and refeathered before the autumn. This was to ensure that they started laying again before the daylight hours dwindled too much. Without artificial light some of the older breeds would not lay at all through the winter, if their moulting was in autumn. The advent of artificial lighting and highly bred hybrids makes such practices less urgent, although most units do try to have all their moults completed before the onset of autumn.

Figure 6.2 The moulting process: the numbers indicate the sequence of feather loss

95

Force moulting, by keeping birds in a hot, dark house, with access only to oats and water for a couple of weeks, is not recommended on humanitarian grounds. What is far more acceptable and effective is to keep the birds confined in their normal house with adequate ventilation, but to provide a maximum of 8 hours of light. During this time, the compound feed should be replaced with oats, but the normal grain ration of wheat should be continued, along with water. There is far less stress on the birds in this way and moulting normally starts a few days after the oats ration is introduced. After 2 weeks' confinement in this way the moult usually proceeds normally and the birds are soon refeathered. It is important to stress that they should be able to exercise, move about and scratch in clean litter during this time. While the birds are so confined, any eggs which are produced cannot be legally sold as 'free-range' ones.

After 2 weeks the lighting pattern is stepped up, with a gradual increase to a maximum of 15 hours. The compound ration is also reintroduced in place of the oats. The flock should be back to full production within a month, continuing until about 100 weeks of age.

During the moulting period a careful watch should be kept for any evidence of stress, feather-pecking or mite infestation, or for any signs of abnormal behaviour. Hanging some greens in the house will provide interest and help to prevent boredom. Some poultry-keepers like to provide a feed supplement such as the traditional 'poultry spice' in the rations. This is basically a mineral and vitamin supplement which helps to ensure speedy refeathering.

REPLACING THE FLOCK

Free-range producers are normally advised to cull flocks at 72 weeks, or to moult them at around 60 weeks for another laying cycle. In other words, the birds are kept through their first laying period, the moult and a subsequent period of lay. Once that comes to an end there is an overall decline in the number of eggs, although the proportion of large ones may be high. Some producers keep their flocks longer, partly because of this factor, but also because they may find it suits their particular management programme. It is something which the individual producer must decide for himself. What needs to be taken into consideration is that, with older birds, there is likely to be a higher incidence of misshapen eggs, and birds with oviduct prolapse problems and broodiness.

There are several options when it comes to dispersal of the flock. Food companies which manufacture poultry products or pet foods may wish to purchase the birds. There are specialist companies which will come to the site and remove the flock. If this is the case, it is important to ensure that the birds are confined to the house, with food and water, to make them easier to round up. 'Cull' birds, as they are called, do represent an important proportion of income, with an average price at the time of writing of 50p per bird. This should not be taken as a static figure, for prices obviously vary. The poultry press normally carries advertisements from those who specialise in the dispersal of poultry flocks.

Domestic poultry-keepers in the area may wish to purchase a few birds for home egg production. This is likely to be a small area of sales, but an advertisement in the local press may prove useful, as long as the purchasers come to collect the birds.

Replacement birds are normally bought as point of lay pullets at around 18 weeks of age. Table birds are either bought as day-olds or as 6-week-olds. It is possible to rear your own replacements, but it is not normally advisable to do so unless all the facilities and buildings for doing so are available. Details of this activity are given in Chapter 9 on Breeding and Rearing.

COPING WITH PREDATORS AND VERMIN

The main enemy of the free-range producer is undoubtedly the fox. There is really only one option available and that is to exclude him entirely from the site. This necessitates high and secure perimeter fencing, as described on page 88. Wandering dogs, which can also be a problem, will be kept out by efficient fencing.

It is not possible to exclude the rat, so regular control to keep down numbers is essential. Watch out for tell-tale signs such as droppings and gnawed areas of woodwork, and apply a proprietary product to the affected areas. The Poultry Laying Flocks (Collection and Handling of Eggs and Control of Vermin) Order, 1989 makes it a mandatory requirement for egg producers to prevent vermin infestation of poultry houses and egg stores. The local authority is in a position to enforce this order.

The presence of
vermin is a health
hazard, as well as
a nuisance.
(Sorex Ltd)

Poison needs to be placed carefully so that only the rats have access to it. It should be quite inaccessible to the chickens, wild birds, domestic pets and, of course, children. Companies which sell rodent control products provide excellent advice and guidance on the placing of rat poison, and their suggestions should be followed. Local authorities have pest control officers whose experience is invaluable in this respect, and their services can also be called upon. Domestic poultry-keepers are normally given this service free of charge, but commercial premises are usually required to pay.

All feedstuffs should be stored in rodent-proof buildings, and it is important not to leave food lying around. Mice can also be a problem, although not to the same extent as rats. The latter are particularly dangerous because they carry Weill's disease, a condition which is potentially lethal in man. Rat urine can transmit it, and it is important not only to clean and disinfect areas regularly, but also to wash your hands thoroughly after handling anything which may have been in contact with rats. Ensure that any cuts on the hands are covered and protected before doing anything.

Mink, which are a problem in some areas, are normally found near lakes and rivers. There is really only one solution and that is to resort to trapping. The traps need to be specific to mink so that other wildlife and domestic pets are not endangered. The Ministry of Agriculture will advise on this, and may lend traps to affected producers.

KEEPING RECORDS

Keeping adequate records is essential in any business. A free-range enterprise is no exception. Even the smallest poultry-keeper who is not necessarily selling eggs or table birds would be wise to do so, otherwise he may find that it is costing him more to produce them than bought-in ones. The salient production records to keep are:

- Number of eggs or table birds produced in relation to number of birds in the original flock.
- Amount of feed consumed.

Egg records

The number of eggs collected should be recorded every day. Remember to make a note of any broken ones. The eggs can either be totalled every week or every 28 days, which will give 13 recording periods in the laying year. These totals will apply to the original number of birds in the flock, so remember to record all mortalities. On a small scale, it helps to have a calendar hanging up in a convenient place, so that the total can be jotted down immediately. Numbers can be transferred to a more permanent record book later. It is easy to think that the totals can be remembered, but they never are. Unless they are written down every day, they will be forgotten!

Larger producers will have recording systems tailored to their own units and possibly computerised records. There is specialised software available which can keep the records of several flocks.

Table bird records

The first thing to record is the number of birds in the batch. This may be a group bought in as day-old chicks, hatched on your own premises or bought in as 6-week-old pullets. Any mortalities should be recorded so that the final production statistics relate to the original number of birds.

Record the date at which batch production commences and, thereafter, the amount of feed consumed every day, or at whatever period is appropriate for your own management system. My own practice was to record every day in a notebook hanging up in the

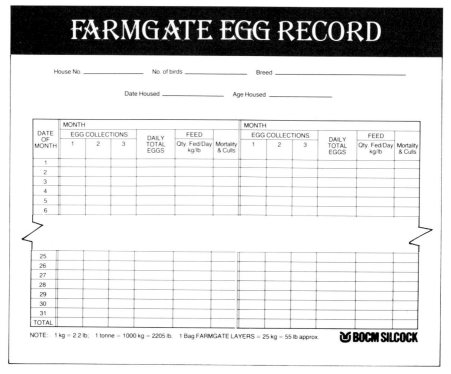

Figure 6.3 Egg production records

feedstore, and to transfer data once a week to a more permanent record book.

It is necessary to keep regular checks on weight so that the amount of weight gain in relation to feed consumed can be calculated. Large broiler units often use electronic weighing systems which will then work out the feed conversion ratio – the average amount of feed consumed in relation to average weight gain. On a small scale, weighing a few birds from each batch, once a week, is usually sufficient. A suspended spring balance with a canvas bag is effective. The canvas keeps the wings confined, while the hole at the bottom allows the bird's head to emerge, keeping stress to a minimum.

The feed conversion ration can be worked out by seeing how much food is eaten for every 450 g (1 lb) weight gain. If, for example, 1.35 kg (3 lb) of food is eaten for every 450 g (1 lb) of weight achieved, the feed conversion would be as follows: FCR = 3:1. The average intensive broiler unit aims for a feed conversion of around 2, with birds reaching a weight of approximately 2.05 kg (4.5 lb) in 45 days. Obviously, such an aim is meaningless to a free-range enterprise where slow growing birds are kept for about 13 weeks, but the general principle of recording grain consumption in relation to weight gain in order to find the feed conversion ratio is just as valid.

A simple recording system which I used for my own table bird enterprise is shown below.

Figure 6.4   A simple recording system for table birds

Date commenced		Comments		
Number in batch				
	Feed consumed	Average weight gain	Mortalities	FCR
Week 1				
Week 2				
Week 3				
etc.				

Note: It is obviously necessary to begin by recording the average weight of the batch if they have been bought in as pullets.

Feed records

On a small scale, it is easy to record every time a sack of feed is opened, as long as there is a convenient hook on the door of the feedstore. When the sack is opened, tear off the label and put it on the hook. When you come to writing down the information in a record book, it is then just a matter of looking on the hook. Be sure that the price of the sack is recorded as well.

Larger producers may find it more difficult to record the amount of feed used in a given time, particularly if bulk deliveries and an automatic feed system are used. The solution here may be to install a feed weigher so that regular checks of feed provision can be made. Feed intake can then be checked manually once a week, by hand-filling suspended feeders or chain feed hoppers and seeing how much is eaten in a given time.

Financial records

It goes without saying that all purchases, transactions and sales should be carefully recorded. The way in which this is done is best organised after consultation with your accountant.

ADAS and several of the feed companies run costings services for the large producer. These schemes enable both performance and cash flow records to be monitored accurately, as well as providing advisory facilities on management programmes.

References

1 Hubbard Poultry UK, press release, 31 January 1986.

7 EGGS

EGG QUALITY

In 1859 Mrs Beeton extolled the virtues of the egg as a 'delicate food, particularly when new laid'. Modern consumers are equally concerned with freshness, but they are also demanding quality and this extends to appearance, colour and freedom from additives and infection.

Grades

The EC regulations governing egg sales define the quality of eggs to aim for by a system of grades. The top quality, and the one to pursue, is Grade A which is legally defined as follows:

Grade A: These are fresh eggs which are normal, clean and un-damaged, and with a stationary air space of a height not exceeding 6 mm.

The white should be clear, limpid, of gelatinous consistency and free of extraneous matters of any kind.

The yolk should be visible on candling as a shadow only, without a clearly discernible outline, not moving appreciably away from the centre of the egg on rotation, and free of extraneous material of any kind.

There should be no perceptible development of the growth cell or extraneous odour, and the eggs should not have been washed or cleaned by any other means.

Grade A eggs should not be refrigerated below 8°C.

There are other grades, but these are not applicable to free-range eggs. Grade B, for example, includes preserved eggs.

Every producer should instigate a 'quality control' programme and examine a representative batch of eggs every so often. This will include 'candling', or examining some eggs against a bright light in a dark room, and breaking some open to check their appearance and consistency, as detailed in the requirements above.

Shell colour

Consumers require free-range eggs to be brown! There is no rational basis for the belief that brown eggs are better than white, but it is widely perceived to be so – and it would be a foolhardy producer who attempted to go against this perception.

The factor for shell colour is primarily a genetic one so it is important to acquire birds which lay brown eggs, not those which lay white ones. This aspect was discussed in Chapter 4, along with details of recommended breeds.

Egg size

Large eggs command the best price and it is important to be able to produce the maximum number of eggs in the top three weight bands. The recognised sizes are as follows:

Size 1 70 g and over
Size 2 65 g and under 70 g
Size 3 60 g and under 65 g
Size 4 55 g and under 60 g
Size 5 50 g and under 55 g
Size 6 45 g and under 50 g
Size 7 Under 45 g

Most stock suppliers will provide an egg weight profile for their birds, indicating the percentage of large eggs which can realistically be expected during the laying period. An example of such a profile is shown on page 66 of Chapter 4.

Feeding naturally plays an important role in egg size and adequate levels of concentrates must be made available. The cost of compound feeds should also be set against the projected sales of particular egg grades. For example, it may be worth feeding extra compound feed to produce the maximum number of size 1 eggs if there is an assured market for them, but if a producer has been asked for a regular consignment of size 3 eggs it would be sensible to cut the compound feed ration slightly and boost the grain allowance. It is one of the most difficult balances to get right and specialist advice from ADAS or the feed company is advisable. In large enterprises, feeding ratios are frequently worked out by computer in order to maintain precise control over a given period. Some feed companies also produce a range of free-range feeds, depending on conditions and requirements. The ration may, for example, have a higher protein content than normal, with extra linoleic acid for large egg production.

Egg shape

Everyone knows the ideal shape for a chicken's egg. Those which do not conform because they are ridged or otherwise misshapen are likely to be viewed with suspicion. There may also be problems fitting them into egg boxes.

The first eggs from early layers may be small and misshapen, but this is fairly uncommon and normally temporary. The real problem emerges with older birds, particularly those which are coming to the end of their laying period, or which are kept for a second year.

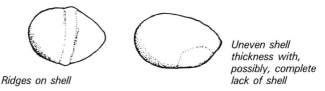

Ridges on shell

Uneven shell thickness with, possibly, complete lack of shell

Figure 7.1   Examples of misshapen eggs

Shell strength and texture

Shell strength is important if a high incidence of cracked eggs is to be avoided. Genetic factors are involved to a certain extent, with some stock having a greater potential for producing stronger shells. It is worth checking the results of the European Laying Trials in this respect; the poultry press normally publishes the results each year. Some agricultural colleges also conduct their own trials to establish relative quality. A recent trial conducted by the West of Scotland Agricultural College, for example, indicated that ISA Brown scored best when it came to shell thickness and the lowest number of seconds, while Ross Brown had the heaviest egg weight and lowest mortality. Hisex Brown produced the highest number of eggs, with the lowest feed intake and feed conversion.[1]

105

Shell strength and texture both deteriorate as the bird gets older and it is unwise to keep old birds in the flock. Feeding also plays a vital role, with an important emphasis on the relative balance of calcium and phosphorus. Research indicates the need for higher levels of calcium and lower levels of phosphorus than those advocated in the past.[2] Concentrate feeds are formulated with these recommendations in mind, and it is not normally necessary to give extra supplies of these minerals. An excess of phosphorus can cause ridging and distortion of the shell. Recent research by Harper Adams Poultry Research Unit indicates that the addition of sodium bicarbonate to the feed ration has a beneficial effect on shell strength.

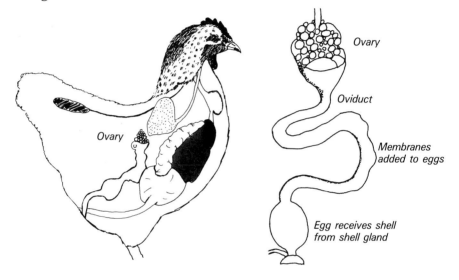

Figure 7.2 The egg-laying system

Yolk colour

The colour of the yolk is important, with consumers rejecting that which is too pale, as well as that which is too orange. The optimum colour range is 11–12 on the Roche scale, a standardised way of determining degrees of yellow-orange. Grass and maize both contribute to a good yolk colour, but most concentrate feeds contain additives to bring about colour enhancement. Consumers have reacted against the use of artificial colorants, so free-range feeds contain natural ingredients such as marigold and capsicum extracts.

Internal quality

Regular testing of a sample of eggs will help to ensure overall quality control. Candling, or examining the internal structure with a bright light source, indicates the height of the albumen in relation to the other cell contents. This can also be established by breaking open some eggs and measuring the albumen height. It is measured in Haugh units and an acceptable measurement would be 80 or above.

Some eggs should also be broken to maintain a visual check for factors such as blood spots or other contamination. If these are discovered, a different management procedure may be called for.

Egg purity

Eggs have a natural and protective bloom which helps to protect them from external contamination. Once they are washed, this protection is destroyed. Free-range eggs which are offered for sale should not be washed; if they are dirty, they are barred from being sold anyway. Every effort should be made to produce clean eggs by maintaining a high standard of management, and by using rollaway nest boxes.

Hands should be washed before eggs are collected. The aim should then be to store them, as quickly as possible, in cool conditions. In this way, the possibility of outside contamination is kept to a minimum.

Internal contamination of the egg by *Salmonella enteritidis* comes from an infected carrier bird which passes on the infection without necessarily being affected itself. This is one source of contamination which the regulations detailed on page 146 seek to control. Wild birds and rats can also be instrumental in introducing it to a site, so it is vital to do everything possible to exclude them.

AVOIDING PROBLEMS WITH EGGS

It is one thing to know the kind of quality egg to aim for; it is quite another to produce it! All sorts of problems can occur unless management is of a high standard. Table 7.1 lists these problems, with details of how to avoid them, as well as how to remedy them if they do appear.

EGG COLLECTION AND STORAGE

Eggs should be collected as frequently as possible, and where they are being sold there is a legal requirement to do so at least once a day (The Poultry Laying Flocks (Collection and Handling of Eggs and Control of Vermin) Order, 1989). This order also states that those collecting eggs should wash their hands before doing so.

On no account should free-range eggs be washed! The aim, as mentioned before, is to produce clean eggs in the first place, and the best way of doing this is to adopt a good level of management and use rollaway nest boxes of the type referred to in Chapter 2 on housing. Washing eggs destroys the bloom of the shell, a natural, protective barrier which helps to prevent the entry of bacteria.

Eggs offered for sale must be in cartons which are appropriately labelled. (BOCM Silcock)

Table 7.1 Problem eggs

Problem	Causes	Remedies
Muddy eggs	Muddy conditions.	Improve area around pophole entrances.
	Dirty litter.	Replace litter regularly.
Blood stains on shells	Young stock coming into lay early.	Do not give too much light, too early, and adjust light intensity with dimmers.
	Overfeeding of protein causing oversized eggs.	Adjust feeding programme.
	Older stock laying oversized eggs.	Do not keep old birds in the flock.
	If blood also in droppings suspect coccidiosis.	Consult vet about prescribing coccidiostat.
Broken or eaten eggs	Too much light early in day.	Adjust the lighting programme.
	Light intensity too great.	Use dimmers.
	Birds underfed.	Adjust feeding programme.
	Easy bird access to eggs.	Use rollaway nest boxes or fit vertically slit plastic curtain.
	Infrequent egg collection.	Collect eggs frequently.
	Bored birds.	Suspend cabbage greens outside for them to peck.
	Inadequate grit available.	Make grit available.
	Thirsty birds.	Check availability of water especially in summer.
Small eggs	Inadequate protein.	Adjust feeding programme.
	Pullets in lay too early.	Avoid giving artificial light too early.
	Hot conditions.	Provide adequate insulation and ventilation in the house.
		Provide shading outside in the summer.
Misshapen eggs	Birds too old.	See previous references.
	Stress and shock.	Avoid overcrowding and sudden noises. Keep dogs away from site.
	Disease, e.g. fowl pest.	Consult vet if problem is more than temporary.
Small number of eggs	Unsuitable breed.	Buy modern, commercial hybrid strains.
	Not enough food.	Feed layer's compound ration with adequate level of protein.
	Coping with winter conditions.	Increase grain ration in cold weather.
	Not enough water.	Improve access to water.
	Not enough light.	Provide extra light as normal daylight dwindles.
	Eggs being eaten.	See references above.
	Eggs being taken.	Check security of house and site.
	Disease.	Consult vet and check vaccination programmes.

109

(continued)

Table 7.1 Problem eggs (continued)

Problem	Causes	Remedies
	Eggs being laid elsewhere.	Buy birds which are floor reared, not cage reared. If necessary, train birds to use nest boxes. Make enough nest boxes available. Check for nests outside and remove any. Place nest boxes below windows, out of the light, and provide adequate perch access to them.
Soft-shelled eggs	Shortage of calcium or too much phosphorus in relation to calcium.	Feed a proper layer's ration where the balance is correct.
	First egg of pullet coming into lay.	Nothing to worry about as long as it does not continue.
	Old stock.	Do not keep birds after two production cycles.
	Disease, e.g. fowl pest.	Consult vet if the problem is more than a temporary one.
Pale yolks	Not enough pigment in feed.	Use free-range layer's feed with natural pigments, or give extra maize.
	If permanent and accompanied by blood in droppings, suspect coccidiosis.	Consult vet and give a coccidiostat.
Greenish yolks	Birds eating weeds such as shepherd's purse or acorns.	Remove plants from grazing area.
Double yolks	Fairly common in large eggs.	Not a problem and often popular with customers.
Egg with no yolk (wind egg)	Fairly uncommon. Usually the result of sudden shock. May occur in a pullet's first egg.	Avoid stress and noise.
Blood spot in egg	Result of blood escaping from ovarian follicle.	More common in old birds. Avoid keeping old stock. Candling and breaking sample eggs will help to maintain egg quality.
Developing embryo	Fertile egg.	Do not let cockerel run with layers.
Parasitic worm in egg	Extremely rare and only in birds with a high burden of worm infestation.	Ensure that free-ranging birds are treated with vermifuge, and that clean pasture is regularly available.
Egg which tastes of fish or garlic	Excess fishmeal in layer's ration.	Check ration or use layer's ration based on plant proteins.
	Wild garlic eaten on pasture.	Check site for plant culprits.

110

Egg trays for storing eggs in cool conditions while awaiting sale.
(Omni-Pac UK Ltd)

There are undoubtedly abuses taking place in this field, with some free-range producers clandestinely washing eggs before selling them. This is highly irresponsible: not only does it pose a health risk to the consumer, but it is also a deliberate fraud perpetrated on those who are paying a premium price for what they believe is a quality product.

The sale of cracked eggs is illegal. They should be used as soon as possible for domestic purposes, ensuring that they are well cooked before consumption.

If eggs are being sold at the farm gate or being supplied to an egg packing station they do not need to be graded and sorted into sizes. (The grades and sizes are the qualities and weight bands referred to on pages 103–4.) If a producer wants to do his own grading and sizing, however, he must not only be registered as a producer, but also as an 'egg packing station'. Details of how to go about this are available from the Egg Marketing Inspectorate of the local authority. Purpose-made egg graders, large and small, are available from specialist poultry equipment suppliers.

Once collected, the eggs should be used or sold as quickly as possible so that they are absolutely fresh. If they have to be stored, it should be in a cool room where the temperature is between 10 and 12°C. On a small scale, this may be in a pantry or outhouse. A larger unit will have its own egg store, well insulated to avoid temperature fluctuations, and equipped with fans which come into operation when the thermostat indicates that the temperature is rising.

Any outside buildings used as egg stores must be protected from rats, and there is a legal requirement to do so if the eggs are being sold.

The eggs can be stored either in egg boxes or in paper fibre Keyes trays, ensuring that they are in a clean, dust-free environment. Details of packing and labelling are to be found in Chapter 10 on Marketing.

References

1 West of Scotland Agricultural College Laying Trial, 1987–8.
2 Gleadthorpe Experimental Husbandry Farm research.

8 TABLE POULTRY

'Foods of the future will have to convey a theme of
fitness and fun, be convenient and of high quality.'
– Colin Groom, ADAS Marketing Adviser, 1986

In Chapter 3, reference was made to the red-feathered broilers,
ISA 657 and Shaver Redbro, which have been developed for non-
intensive conditions. They are more slow growing than the white-
feathered strains such as Cobb and Hubbard broilers, but are well
adapted to outside conditions. In France, where the strains were
originally developed, the *Label Rouge* or Red Label sector represents
15% of the total table poultry market, with over 50 million birds
produced each year. Anyone considering a table poultry enterprise
would do well to concentrate on this quality sector which is still in
its infancy in Britain.

The choice of birds could also include the white-feathered strains
referred to above. Although the Cobb has been bred for intensive
conditions, in my experience it adapts well to extensive conditions,
and the exercise it has on range ensures that leg weakness problems
from over-rapid growth do not materialise. The traditional table
breed Light Sussex may be the choice of some small poultry-keepers
who have a particular interest in pure breeds.

In intensive conditions, the aim is to produce a marketable bird
as quickly as possible, with an important emphasis on rapid growth
and efficient feed conversion. In the extensive sector, birds are
raised for a minimum of 81 days before slaughter, with 13 weeks
being closer to the average. During this time, they are housed in
similar conditions to free-range layers, with access to pasture range,
although they may be confined for the last couple of weeks for the
finishing stage. Their diet is predominantly a grain one, and the
so-called 'corn-fed' birds attain a yellowish tinge to the skin from
the maize (sweetcorn) in their diet. In France, many of the Red Label
producers grow their own maize/sweetcorn, thus keeping down
overall feed costs.

A white Cobb broiler. Although bred for intensive rearing, the Cobb adapts well to free-range conditions, as shown below.
(The Real Meat Company)

The highest return on table birds is on those which are 'organically produced', but producers need to be registered as organic producers and the requirements are often difficult to meet. National standards for organic production have been drawn up by the United Kingdom Register of Organic Food Standards (UKROFS). Producers can apply for registration and for the right to use the UKROFS logo on their products. In order to meet the requirements, a producer must conform to the following:

- No prohibited substance should have been applied to grassland in the previous 6 months.
- Poultry for meat production must be brought in as day-old chicks, unless bred on the site.
- As far as possible, systems for producing poultry should be based on grazing, but they may be finished indoors.
- Housing and management must be appropriate to the behavioural needs of the birds.

114

- The feed should not contain any prohibited substance.
- At least 70% of the ration should be obtained from UKROFS registered and approved sources.
- Any medications should be confined to the essential health and welfare of the bird, and should be recorded.

Registered producers who abide by organic standards may use the recognised UKROFS logo.

There are purpose-made feeds and grains available from specialist feed suppliers which meet the organic standards requirements, but they are still expensive and in relatively short supply in Britain. If a producer is able to grow his own grain to the required standards, it becomes a more economic proposition. Full details of specific requirements may be obtained by contacting UKROFS.

Many producers experience problems in meeting the standards for organic production and so concentrate on producing table poultry as non-intensively as is possible in their circumstances. This is usually with the same general considerations as those outlined for organic production, but the 70% grain ration is not grown to organic standards. Organically grown grain is nearly double the price of normal grain. Even so, the returns are high compared with those for intensively reared birds. Table 8.1 shows the relative difference in ex-farm and retail prices for the three production methods. They are average figures for the country as a whole and should not be taken as typical, for prices paid by individual packers vary considerably.

Table 8.1 Comparison of returns on intensive, non-intensive and organic table birds

	Wholesale price (£) (live weight per lb ex-farm)	Retail price (£) (fresh/oven-ready per lb)
Intensive poultry	0.25	0.98
Non-intensive poultry	0.55	1.75
Organic poultry	1.00	2.50

Source: author's costings, July 1989.

115

MANAGEMENT OF TABLE POULTRY

In order to meet the requirements of the organic standards, table birds must be brought in as day-old chicks, unless they are being bred on site. An alternative is to purchase them at around 6 weeks, when they no longer require heated conditions, and rear them as non-intensive birds. In Chapter 9 on Breeding and Rearing, full details of protected rearing are given.

Once they are off heat, the young birds can be housed in build-ings similar to those used for free-range layers, including black polythene housing if this is available. They need to be protected against foxes, and electric fencing is the most effective means of doing this.

My own system of housing table birds is indicated in Figure 8.1. This includes an ordinary perchery house of the kind that can be moved from one area to the next without too much difficulty. A chicken wire fence was erected in front of the house to make a straw yard and the birds were confined to this area if weather conditions were poor. In fine weather the gate was left open so that they could range on grass in the rest of the field but, as I did not have electric fencing for a number of years, considerable care had to be exercised in relation to foxes. One year, the fox came before dark and I lost nearly a third of the batch.

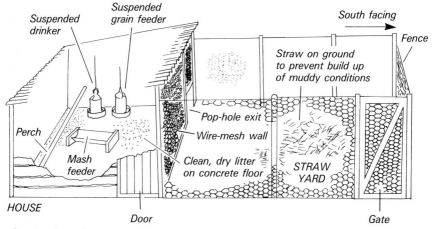

Source: the author

Figure 8.1 Example of a small free-range table bird management system

116

The predominantly grain ration of table birds can be made available on an *ad lib* basis in the house, although when I visited a farm in France I was interested to see that their food was put in large outside hoppers. These were placed all over the site to encourage wide ranging, and the birds were fed exclusively on maize (sweetcorn) grown on the farm. This was chopped up using an electric grain grinder equipped with kibbling plates so that the resulting pieces were of an appropriate size for the birds.

For their first 4 weeks, chicks will need to be fed on a starter ration of chick crumbs. Most of those available contain coccidiostats for the routine control of coccidiosis, but it is now possible to obtain them without additives. Similarly, grower's concentrates are now available without the range of antibiotics and coccidiostats that they previously contained as a matter of course. These can be given from the age of 4 weeks, but as they are usually 19% protein for rapid growth, many producers prefer to restrict them to 30% of the total ration, with the bulk consisting of grain. As referred to earlier, some producers rear their table birds exclusively on grain from around 4–6 weeks. Others feed grain from this age until 2 weeks before slaughter, when the birds are housed for finishing. At this point, they provide additive-free grower's pellets as 30% of the total ration.

Water and insoluble grit should be made available to the birds at all times, with shading being provided for the outside drinkers in hot weather.

The age at which to slaughter the birds depends on marketing strategy. If they are being reared to contract, they will usually be collected from the site when the distributor specifies. For the small producer the ideal situation is for the butcher to collect the birds live from the farm, but this is by no means guaranteed, and the producer may have to slaughter and pluck the birds himself.

It is vital to have an assured market before starting an enterprise. There is nothing worse than rearing the birds and then wondering how to dispose of them. It is highly recommended that anyone thinking of raising table poultry commercially should attend a course such as that run by the Agricultural Training Board or one of the agricultural colleges. Management and marketing are normally covered in such courses, and there is usually an opportunity to undertake practical work at existing units.

There are companies who specialise in marketing and distributing organically or non-intensively reared table poultry, as well as those who specialise in distributing organic produce. They are listed in the Reference section at the end of the book.

The regulations on testing poultry flocks for salmonella on a regular basis do not apply to table birds, as long as they are reared only for this purpose. If they are reared or sold as potential breeding stock or egg producers, they will need to be tested. In all cases, every care should obviously be taken to avoid contamination and to ensure hygienic standards. This entails buying in chicks or young birds which are guaranteed to be healthy and free from infection, maintaining scrupulous standards of cleanliness in housing and management, and ensuring that no questionable food is given to the birds. Kitchen scraps, for example, are definitely out if the birds are to be sold. It is now possible to buy free-range table bird feeds which are based on grains and vegetable proteins, and which are free of many of the additives which feed used to contain. Feed compounders are also now subject to much stricter legislation, to ensure that feed components are not contaminated with micro-organisms such as salmonella bacteria.

Slaughtering

As mentioned earlier, it is much better for the producer to raise the birds on a contract basis, if possible. This means that he is not involved in the slaughtering and processing because the distributor will collect the birds from the site and deal with them. Obviously the numbers involved would need to be fairly substantial for this to be a possibility, although local butchers, for example, may agree to take far smaller numbers on a regular basis.

If the birds are to be slaughtered on site, they should be housed in protected conditions and have access to fresh water at all times. Grain should not be given for 24 hours, but dry mash or concentrate pellets can be given until 6 hours before slaughter. Such a practice will ensure that the gut is relatively clear, without causing stress to the birds.

The normal practice is to use a humane killer so that death is instantaneous. On a large scale, this would involve the use of an electric stunner, followed by piercing of the jugular vein to bleed the bird. On a small scale, there are now several poultry killers which cut off the head in a swift, guillotine action. It is important to stress that slaughtering should only be undertaken by an experienced person. Those who are inexperienced, but psychologically able to

Humane poultry killer.
(Warren Agricultural Machinery)

undertake the task, will find it useful to go on a practical course of the type referred to earlier.

Some butchers may require poultry which has been slaughtered and plucked but not gutted, and with the head and feet left on. If this is the case, the slaughtering will need to be by neck dislocation rather than decapitation.

Killing by instantaneous neck dislocation.

119

Plucking

Freshly killed, unplucked birds may be sold in local markets and butchers, but there is little doubt that most people prefer to buy a plucked and eviscerated bird.

There are three methods of plucking – traditional hand plucking, machine plucking and wax plucking. The first method is still widely used by small farmers, particularly those who are catering for the Christmas market. At this time, it is common to employ part-time, seasonal labour.

It is a good idea to wear an overall for hand plucking. Those with allergies are advised to wear a face mask. A large, airy barn, out of the wind and with a concrete floor, is the ideal plucking area. The feathers will not blow about and are easily cleared from the floor area.

The feathers will come out more easily while the carcass is still warm. The skill lies in removing the feathers without tearing the skin. The sequence will be a matter of personal preference. Some people start with the breast feathers, then go on to the back, sides and, finally, the wings.

On a larger scale it may be more appropriate to invest in a plucking machine. These come in a range of sizes. Waxing is another possibility. Here the bird is immersed in a wax bath and, as it cools, the wax sets, allowing the feathers to come away with the wax when it is pulled off. Several other proprietary products are available and these are listed in the Reference section at the end of the book.

Processing poultry

A good, sturdy beech table is ideal for this stage, but any surface which can be scrubbed after use is suitable. Have plenty of muslin cloths for swabbing soiled areas, and buckets or large collecting utensils for the entrails. A large enterprise will have purpose-made premises, but a small farm can use a cool, airy outhouse to good effect.

Cut off the head and make a cut in the skin along the back of the neck. Use poultry secateurs to cut through the neck and remove it. A piece of muslin placed around the neck will make it less slippery to get hold of. Enlarge the incision in order to get the hand in after the neck has been removed, and take out the gullet. Now fold over the flap of skin and twist over the wings so that they hold it in place.

Turn the bird around and make a circular incision around the vent so that it can be pulled clear. Take care not to pierce the rectum. As the vent is pulled clear, the intestines follow after and can be dropped into a bucket. Enlarge the opening enough to get your hand in and draw out the gizzard, liver, crop and lungs. The liver, heart, neck and gizzard can be retained, for many customers like to have these for making stock or paté. The gizzard will need to be cut open and the stones and tough membrane discarded. The giblets can then be put in a plastic bag ready for placing inside the carcass when everything else is finished.

Now make an incision just above each foot but do not cut right through. Break the leg bones by snapping them over the edge of a table and the leg tendons will be revealed. They look like white elastic bands and need to be removed because they are very tough. The easiest way of doing this is to have a purpose-made tendon remover screwed into the wall. This is rather like a double hook and when the foot is placed in between the prongs, and the carcass is drawn down sharply, the tendons are pulled out. Tendon removers are available from poultry equipment and farm suppliers.

The bird should now be stored, breast side up, in a cool room ready for sale or use.

Where the enterprise is a large one, it is a good idea to obtain ADAS and local authority advice before any buildings are erected or

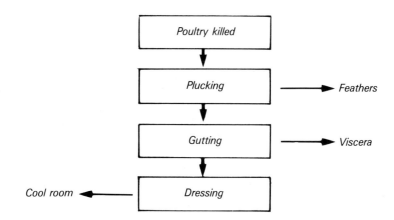

Figure 8.2 Flow chart for a table bird production system

121

adapted as processing areas. They will provide copies of the appropriate food legislation which applies under these circumstances.

Avoidance of cross-contamination

Hygienic conditions are essential at all times and obviously every effort should be made to avoid cross-contamination during the processing of poultry. Clean overalls should be worn and there should be access to hand-washing facilities at all times. The flow chart below indicates a system where the initial plucking and gutting operations are kept separate from the subsequent dressing of the birds. Feathers and viscera should be disposed of immediately, either by incineration or by removal by a contract firm if the scale warrants it.

9 REARING AND BREEDING

This chapter is relevant to anyone who rears young poultry, whether it be day-old chicks as replacement laying pullets, pullets for sale or young table stock. It also covers breeding and incubation practices for those who wish to concentrate on pure breed or first cross production.

REARING DAY-OLD CHICKS

One of the many branches of the poultry industry is that of pullet rearing, where day-old chicks are bought from one of the breeding companies and raised until they are anything from 12–20 weeks old. At this stage, depending on the needs of the buyer, they are sold

Interior of a large rearing house with newly hatched chicks which will be reared to 18 weeks old. (Mick Dennett. Bibby Agriculture)

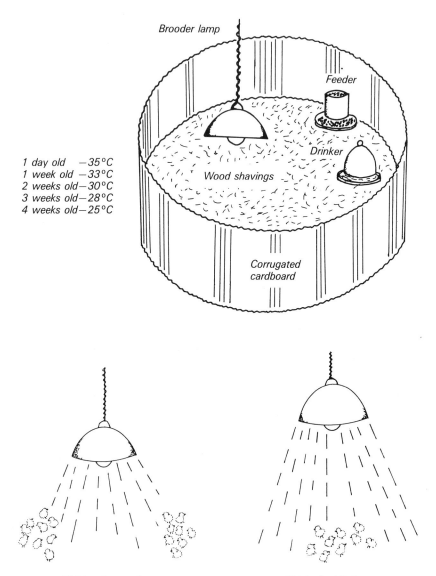

1 day old   −35°C
1 week old −33°C
2 weeks old−30°C
3 weeks old−28°C
4 weeks old−25°C

Brooder lamp

Feeder

Drinker

Wood shavings

Corrugated cardboard

Chicks dispersed round edges—too hot. Raise lamp

Chicks huddled in centre—too cold. Lower lamp

Figure 9.1   Rearing day-old chicks

124

as pullets coming up to lay. They are usually bought at 18 weeks, but there are some buyers who prefer to acquire their replacements early, to ensure that they settle in well before the point of lay period is actually reached. Replacement pullet rearers normally come to an arrangement with a breeding company to act as a local agent for that particular breed.

There may be those who buy day-olds as their own laying replacements, and not for resale to others. This is certainly cheaper than buying point of lay pullets, but it does mean that housing and brooding equipment must be available for them. Not all free-range farms have the necessary buildings and equipment, and the management system may not be geared to this type of diversification.

There are those who buy day-old chicks for table rearing, as discussed in Chapter 8. The aims may be different from those of the egg producer, but the initial rearing management of day-olds does not vary a great deal. Nor does it vary with numbers. The general principles of rearing 5,000 chicks are the same as for half a dozen. They all have the same requirements of shelter, warmth, food and water. It is merely the economics of scale which differ.

A poultry house or outbuilding is required, ideally with a solid floor so that wood shavings can be utilised as litter. A polythene house of the type discussed earlier in the book can also be used, as long as measures are taken to exclude rats which will kill chicks in large numbers.

There is a variety of heaters which can be used to provide warmth for the initial period, powered by propane, electricity or paraffin. One of the most popular methods of providing artificial warmth is with radiant lamps suspended above the ground. It is easy to establish the correct height! If the chicks tend to huddle in a tight mass underneath, they are cold and the heater needs to be lowered. If they are scattered around the periphery of the lamp's output, they are too hot and it should be raised.

With relatively small numbers, it is a good idea to use corrugated cardboard to make a circular shelter, to confine them in a small space initially. As they grow bigger and more adventurous it can be discarded, allowing them to move around the rest of the building. There is also a range of purpose-made brooders available from suppliers. These are self-contained units which incorporate a heating element, and in which the chicks stay until they are able to do without artificial heat.

Suspended drinkers are best, ensuring that water is not splashed

Brooders for raising
young stock in
protected conditions:
gas-powered (left)
and electric (below).
(Maywick (Hanningfield)
Ltd)

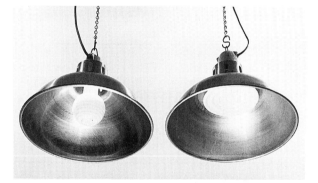

about, while chick crumbs can be made available either in suspended, gravity-fed feeders or shallow pan containers. If the latter
are used, it is important to ensure that they do not become fouled
with droppings. Most people find that the best policy is to follow
an *ad lib* pattern of feeding, where the feed is available at all times.
As the young birds grow, they can be gradually switched to an
additive-free grower's ration, together with grain.

Wood shavings provide a warm, effective litter layer for the
chicks to move around in, although chopped straw and shredded
paper are increasingly used. My own choice is wood shavings, but
if there is a cost-effective local source of either of the other two,
it makes sense to consider these alternatives. The point has been
made earlier that free-ranging birds need to be ground reared from
the start, so that there are no problems with moving about and
perching. Regular checks should be made on the condition of the
litter to ensure that it does not become wet or smell of ammonia,

126

two situations which can lead to disease problems such as coccidio-sis or respiratory difficulties. Wet litter can also cause foot problems: it adheres to the claws, gradually solidifying into concrete-like balls. Hock burn is a problem in intensive broiler houses, where the wet litter causes inflamed patches on the legs, but it is rare to find it in less intensively run units.

And what of lighting? If the house has windows, the natural light coming in will be sufficient for the chicks. I am not in favour of the large, barrack-like, environmentally controlled intensive houses, where no natural light enters and a dim, reddish glow like some-thing from a distant inferno is all that is available to the chicks. Again, it should be stressed that these birds are to be free-ranging ones as soon as possible. Where natural light is available it should not be supplemented by artificial light, otherwise early laying may begin before the body frame is sufficiently developed for outside conditions. If the natural light is too bright, it may be necessary to provide some shading for the same reason.

As the chicks grow, the heat can gradually be decreased. The rate at which this happens will depend upon a number of factors, including the outside temperature and the degree of relative hardi-ness of the young stock. It is not battery cage or broiler house stock that we are involved with here; the birds are future free-rangers and the sooner they are able to experience outside conditions the better. This does not mean going to extremes, but on warm days, where the chill factor is fairly low, the pop-hole exits can be left open for them to explore a well-protected and confined area of pasture. Alternatively, a traditional ark or fold unit can be used.

During the first few days of life, chicks are usually injected against Marek's disease. Other vaccinations to protect pullets include those for Newcastle disease, infectious bronchitis and endemic trem-ors. Consult your veterinary surgeon or ADAS representative for advice, particularly where local conditions may be especially con-ducive to some diseases. Gumboro disease can be a problem in intensive broiler houses, although it has not yet proved to be a major problem in the free-range sector.

Vaccination of poultry is a technique which can be learnt at a poultry management course organised by the Agricultural Training Board or an agricultural college. It is a technique which must be exercised with care, particularly since the incidence of accidental self-injection is comparatively high.

Salmonella testing will also be required where more than 24 birds are involved and the eggs from the birds are to be sold. Details are given in Chapter 11 on Health.

PULLETS

Once the pullets are well grown, they will follow different systems of management, depending upon circumstance. Laying pullets will either be sold to egg producers or transferred to the owner's laying house at around 18 weeks of age. From 20 weeks they can be given artificial light to supplement natural light and encourage them to start laying, although this depends upon the time of year. Table birds may be sold at around 6 weeks to those who wish to rear them to full weight, or kept as free-range broilers on the original site. Pure breeds are often sold as breeding trios of one male and two females at any age from 6 weeks onwards. The male should not be too closely related to the females, otherwise the close inbreeding may result in a higher than normal incidence of deformities.

COCKERELS

It is a fact of life that most cockerels are unwanted. The point is obvious in relation to the raising of laying stock, and many of the larger enterprises either dispose of day-old males or sell them to zoos which use them as food for inmates such as large birds of prey. Where they are disposed of on site, there are humane considerations to be borne in mind, and the Ministry of Agriculture provides guidelines in this respect. A leaflet is available from them, on request.

Cockerels of the broiler breeds can obviously be raised for the table. When day-olds are bought from breeding companies for this purpose, it is normal to specify whether 'all females' (A/F) are required, or whether the batch should be 'as hatched' (A/H). The latter are usually cheaper. They can be raised together unless obvious problems of bullying or sexual precociousness begin to manifest themselves. If this happens, it is a good idea to separate the sexes. With intensively raised broilers the raising period before slaughter is usually so short that they are killed before this happens, but with free-range table birds which are raised for longer periods the problem may manifest itself.

128

Cockerels should not be allowed to run with laying hens, unless fertile eggs are specifically required for breeding.

Where pure breeds or first crosses are concerned, some cockerels may be kept and reared as breeding stock. A breeding male needs to be in first-class condition and to have characteristics which are particularly valuable. He may, for example, come from a line of excellent layers and would therefore tend to pass on this tendency to his daughters. It may be that his plumage, carriage and markings are excellent examples of the standards for a particular pure breed. Whatever qualities are looked for in potential breeding stock, cockerels must be in excellent health and be free of diseases which are capable of being transmitted genetically. Any breeding bird, male or female, should be blood tested for such conditions, particularly *Salmonella enteritidis* which can reside in the sexual tracts, and Marek's disease.

BREEDING

Breeding from hybrids is not a practical proposition for the average poultry-keeper, because the progeny will not breed true and the results will be unpredictable. There are those who will wish to experiment anyway, and there is nothing wrong with that, although it often proves difficult to obtain a good male because the breeding companies, understandably enough, protect their own interests in this respect.

Breeding pure breeds or first crosses is a different proposition, and it is an activity in which the smaller poultry-keeper has always had an interest.

Pure breeds

Most poultry-keepers breed pure breeds because they have a real interest in a particular breed and, again, it is worth pointing out that had it not been for the devoted activities of small poultry-keepers in this respect, some of the old breeds would have died out. At the same time, it should be stressed that sometimes there is too great an emphasis on appearance, and not enough on utilitarian factors such as fertility and vigour. For example, far too much importance has been placed on the need for a 'good' Sebright bantam male to have a square tail. Males with pointed tails have tended not to be used, despite the fact that such birds appear to have more sexual vigour. The result is that Sebrights, in the United Kingdom at least, are difficult to breed from successfully, and there is a distinct shortage of vigorous blood lines.

If females need to be selected for breeding on the basis of their egg laying capacities, it will be necessary to identify their eggs. The best way of doing this is to use a system of trap-nesting. This is essentially a nest box which allows the bird access, but once inside she cannot get out. Where trap-nests are used, they should be inspected very frequently so that confinement is kept to a minimum. The birds will also need to be identifiable and the easiest way of achieving this is by using leg rings. These are widely available from poultry suppliers. Needless to say, any chickens being tested in this way should be housed separately from the normal laying flock so that their routine is not upset.

Breeding cocks should also be identified with a leg ring and given their own quarters. The only time they should be allowed to run with laying hens is when mating is required. There is no truth in the old, rather chauvinistic, belief that hens will only lay well when there is a cock with them. The opposite is often true! Hens which are continually being jumped on may suffer stress and

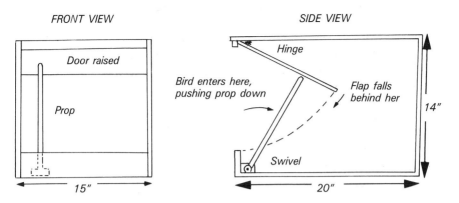

FRONT VIEW   SIDE VIEW

Door raised

Prop

Bird enters here,
pushing prop down

Hinge

Flap falls
behind her

14"

Swivel

15"

20"

Source: Katie Thear. Incubation: A Guide to Hatching and Rearing.
Broad Leys Publishing.

Figure 9.2   Trap-nesting

damage to the back of the neck and the sides. Poultry venereal
diseases such as vent gleet may result, and the eggs themselves
may be contaminated. Where eggs are sold, it is absolutely vital to
ensure that there is no possibility of a fertile egg being included in a
carton. It has happened! In 1989, I purchased a carton of free-range
eggs at a local supermarket, one of several batches in different
areas, as part of the research for this book. It came from one of the
biggest distributors in the free-range field, yet one egg contained an
embryo which was at least 10 days developed, and half filling the
shell. Such an incident is disgraceful and likely to do a great deal of
damage to those who are working hard to further the development
of commercial free-range poultry farming. Breeding cocks must be
kept quite separate from laying stock!

First crosses and autosexing

A first cross is produced when two pure breeds are crossed, and
the resulting progeny inherits characteristics from both parents.
Reference has already been made to the popularity of dual-purpose
breeds in the past, where a bird with a good laying capacity might
be crossed with a heavier, table breed so that the offspring would
initially act as layers, then be used for the table when they came
to the end of their laying lives. The Rhode Island Red male was
frequently mated with the heavier Light Sussex for this purpose.
Another attraction of this particular cross was the ability to be able
to distinguish between males and females at the day-old stage. The

males are silvery yellow, while the females are brownish yellow.

There has been considerable research into the development of auto-sexing breeds, particularly with the barred breeds, where female chicks could be easily identified by the barring across the back. It was the diversification of the poultry industry into the laying and broiler sectors, together with increased efficiency of other sexing methods in the 1950s and 1960s, which led to the decline of research in this area. Hopefully, one effect of the resurgence of commercial free-range production will be to continue with research and development of autosexing breeds for the production of first crosses. Some specialist breeders have continued with this line and there are examples such as the Legbar from a Leghorn cross, a Dorbar from a Dorking cross and a Welbar from a Welsummer cross. Table 9.1 indicates other breed crossings which can be identified as male and female at the day-old stage.

INCUBATION

There are two choices when it comes to incubation – relying on broody hens and using artificial incubators. The first option is obvi-ously only applicable to the small scale, but it is a popular method with small poultry breeders, and the pleasure of seeing a hen with her newly hatched brood is appreciated by many.

A broody hen sitting on a clutch of fertile eggs.

9 REARING AND BREEDING

Table 9.1 Autosexing crosses

Mating any of these 'gold' males with any of the 'silver' females will produce male chicks which are light coloured and females which are buff coloured

Males		Females
Barnevelder	Mahogany Orloff	Light Brahma
Brown Leghorn	Gold Campin	Light Sussex
Brown Sussex	Gold Wyandotte	Columbian Wyandotte
Brown Buttercup	Gold Hamburgh	Silver Grey Dorking
Golden Duckwing	Golden Buttercup	Duckwing Game
Leghorn	Buff Leghorn	Silver Duckwing Leghorn
Black Red Game	Buff Rock	Silver Campine
Indian Game	Buff Orpington	Silver Hamburgh
Partridge Cochin	Rhode Island Red	Silver Buttercup
Black Red Malay	Red Sussex	Silver Wyandotte
Partridge Wyandotte	Red Dorking	White Wyandotte
Wheaten Marsh Daisy		Ancona

Crossing any of the following 'black' males with any of the 'barred' females will produce black male chicks with a light patch on the head (and possibly on the rump) while the females are all black

Males		Females
Australorp	Minorca Spanish	Cuckoo Leghorn
Black Rock	Black Wyandotte	Coucou de Malines
Croad Langshan	Black Cochin	Cuckoo Dumpy
Black Leghorn	Creve Coeur	Barred Rock
Black Langshan	Black Scots Dumpy	Scots Grey
Black Minorca	Black Hamburgh	Cuckoo Maran
Black Orpington	Black Orloff	
Black Silkie		

Crossing these 'dark legged' males with any of the following 'light legged' females will produce male chicks with light legs, while the females are dark legged

Males	Females
Australorp	Barred Rock
Black Minorca	Buff Rock
Brown Leghorn	Brown Leghorn
Black Bresse	Black Leghorn
White Bresse	Buff Leghorn
Buttercup	Light Sussex
Campine	Scots Grey
Hamburgh	White Wyandotte
Langshan	

Crossing a 'dark eyed' male with a 'light eyed' female will produce dark eyed females and light eyed males

Males	Females
Langshan	Brown Leghorn

Source: Katie Thear, *Incubation. A Guide to Hatching and Rearing*, Broad Leys Publishing, 1987.

133

The problem with using broody hens is that they are often not broody and prepared to sit on a clutch of eggs when they are ready for incubation. One way of achieving this state is to give the hen some infertile eggs to sit on in a sheltered coop, in order to trigger off the broody syndrome. Once she is well and truly broody, fertile eggs can be substituted.

Broodiness is easily recognised by the hen's tendency to sit tight and fluff up her feathers at anyone's approach. The breast area is also extremely hot to the touch.

Once the eggs have been substituted and she has accepted them, you can leave her to get on with it, apart from ensuring that she emerges for food and water at least once a day. Normally she will tend to do this at the same time each day, but it is a good idea to place a drinker of fresh water near her at all times.

Once the eggs have hatched, they can be left with her in a protected coop and run so that they are not likely to be bothered by vermin. Alternatively, the chicks can be reared in a brooder such as that described earlier in the chapter.

For most purposes, and certainly for a commercial unit, an artificial incubator is essential. They come in all sizes, from the 25-egg variety up to walk-in ones which can hatch several thousand eggs at a time. The choice will obviously depend on the scale of activities.

There are other aspects to be borne in mind, particularly the need to use the latest technology, even on a small scale. Modern incubators are highly sophisticated pieces of equipment, with automatic egg turning facilities and the use of electronically controlled

A small protected unit suitable for a broody hen and chicks.

134

thermostats. Both of these features are highly desirable, and are to be recommended over the more traditional models. Even small incubators for the hobbyist are now available with these features.

The instructions of the manufacturer should be followed for each model. These will relate to temperature, humidity and egg turning – the three crucial aspects of incubation. Most modern incubators will maintain a precise control over these factors, as long as the instructions are heeded.

Once the chicks have hatched, at around 21 days, they should be reared in the protective conditions described earlier.

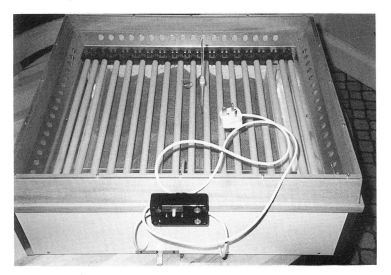

Many of the smallest incubators are now equipped with automatic egg turning facilities. (Brinsea)

135

10 MARKETING

'If you can't identify your customers, forget it!'

It is a truism to say that the market comes before the chicken or the egg but, like all truisms, it contains a fundamental truth. If eggs or table birds are to be sold from a free-range enterprise, their distribution should be researched and established *before* production begins. Similarly, anyone contemplating a small domestic unit should be clear about the aims, and whether keeping pure breeds as a hobby, or producing household eggs and table birds for the home, are realistic activities in relation to time, energy and costs. The key questions posed in Chapter 1 are also highly relevant to this part of the book, and it is recommended that they be read again in conjunction with this chapter.

THE MARKETING PLAN

The techniques of marketing are essentially the same whatever the product. It is necessary to identify the customer and produce a marketing plan for the product. The strategy is outlined in Figure 10.1.

Identifying the customer

I made the point earlier in the book that organisations such as ADAS offer a consultancy service which includes the necessary market research for a particular area. Attending a course on free-range production of the type run by ATB or an agricultural college is highly recommended too. There is also a great deal that the individual can do to identify the potential customer and his attitudes. On a local basis, find out from delicatessens, restaurants, butchers, hotels and guest houses whether they are interested in buying. Speak to local consumer groups, Women's Institutes and environ-

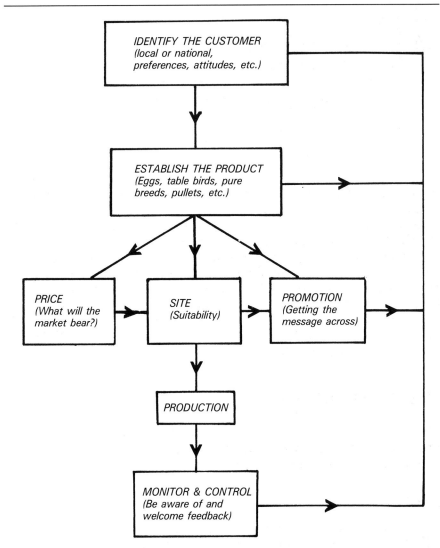

Figure 10.1 The marketing plan

mental groups, and listen to what they have to say. If you are already in production in a small way, try taking a market stall, not only to establish buying patterns, but also to have the opportunity of talking to people. A great deal can be learnt by good listening.

Once satisfied that there is a demand, the decision may be whether

137

to concentrate on local, farm-shop sales, or to cater for national demand via a distributor. The two activities are not mutually exclusive, but the former will obviously require far less capital expenditure than the latter. Most free-range producers are small, catering for local sales, but most free-range eggs come from large enterprises selling to chain retail outlets. If the enterprise is to be a large one, a priority will be to contact one of the major distributors or retail chain buyers and come to an agreement with them. Most of these will require an inspection of the premises and a guarantee that production will be to their standards.

Establishing the product

Establishing the specific product will be largely determined by the initial identification of the customer. There is not much point in launching into egg production if all the market research has indicated that there is a local glut of free-range eggs in the supermarkets, and what customers are looking for is a source of free-range table poultry. What is needed now is a more detailed examination of the specific product. For example, if there is a demand for free-range eggs, is it better to concentrate on the production of large sized or dark brown speckled ones such as those from the Maran breed? There may be fewer eggs but the higher returns may make this worthwhile. If the product is a free-range table bird, is it more appropriate to concentrate on meeting the organic standards, or merely those of general, non-intensive standards? Again, some test marketing in the form of initial small-scale production and local sales may indicate the path to take.

The price

The price of the product is crucial. Mr Micawber, in *David Copperfield*, sums up the recipe for economic happiness as 'Income £1, expenses 19s 6d' – but the aim should not be just to cover costs and have a bit of profit at the end. The positive approach is to ask the question, 'What price will the market bear?' This is the starting point with any quality product, especially where production costs are higher anyway. At the time of writing, the average price of free-range eggs is 35p a dozen above that of battery produced eggs,[1] while that of non-intensively reared table birds is approximately double that of intensively raised broilers. These are averages for

138

the whole of the country, and do not take local variations into consideration.

Keeping in touch with market information is vital and there are several organisations which issue regular details of price changes. One of them is UKEPRA (referred to above) which issues information to its subscribing members. MAFF also has a weekly statistical bulletin on the poultry industry. Local National Farmers' Union (NFU) offices produce a weekly update of pricing information available to their members, and the poultry press also publishes regular updates.

In an affluent area, where there is an increased awareness of, and distaste for factory farming methods, the acceptable premium is likely to be higher than in other regions. Again, it is a factor which will need to be carefully researched before any decision is made.

The site

The site where production is taking place is crucial if this is also to be the area of sales distribution. If an expensive, quality product is being produced, it is not likely to achieve its maximum sales potential if it is in an isolated area. A successful farm shop needs to be close to a fairly highly populated, reasonably affluent and mobile centre of population. Proximity to an urban area may not necessarily lead to successful sales if it is also a region of high unemployment. In such areas, the cost factor is likely to take precedence over considerations of quality.

The ideal road for a sufficiently high volume of calling customers at a farm shop is a two-way 'A' road which carries a substantial proportion of commuter traffic to a nearby city. A straight road, offering plenty of areas for warning notices of the shop ahead, enables drivers to see clearly, and to have sufficient time to respond. There should be plenty of turning and parking space, without the danger of causing a traffic hazard.

Conditions such as these enable some producers to do without a distributor and gain the maximum return on their eggs by selling direct to the public.

Any changes, such as the creation of a dual-carriageway, could be disastrous to an enterprise, not only by halving the numbers able to call, but also by creating a situation where even those on the correct side of the road are discouraged from stopping by the speed of the through traffic. Direct gate sales are the 'bread and butter' of many producers, providing an effective security against the fluctuations of packers' prices, but it is essential to check on future transport plans in the area before going into business.

Promotion

Producing a marketable product is one thing, but promoting it is another. The message has to be got across to customers that here is something which is clean, wholesome, environmentally and ethically sound, and definitely good for you.

It is a good idea to have the marketing advantages of free-range eggs and table birds clear in one's mind. The point has already been made that most people who buy eggs, for example, buy battery ones because they are cheaper. The marketing of free-range eggs is firmly targeted at those consumers who demand a 'quality' product. The immediate advantage of free-range is that it is perceived to be more 'natural' and that there is a greater concern for humanitarian and welfare considerations. This distinction would be even more apparent if intensive producers were required by law to label their eggs as 'battery-produced'. Unfortunately, they are not required to do so, nor are they prohibited from describing their eggs in such terms as 'country produce fresh from the farm', when the reality is that they are 'fresh from the industrial unit of the battery'.

There are some feed producers who produce special free-range feeds without the range of additives which are found in many other commercial feeds. They have a range of marketing aids such as posters, displays and the feed company logo which egg producers can use, as long as they abide by the company's quality requirements in addition to the normal legal needs. These can be a positive advantage when it comes to marketing quality free-range eggs or table birds.

On the question of comparative quality, there is some evidence that the free-range egg is nutritionally superior to the battery egg. The *British Journal of Nutrition* (March 1974) reported that battery eggs were 70% lower in vitamin B_{12} and 50% lower in folic acid. The Ministry of Agriculture subsequently disputed these figures, claiming that battery eggs are 40% lower in vitamin B_{12} and 30% lower in folic acid (1978).

In 1987, the Department of Catering and Domestic Studies at Stafford College of Further Education conducted some egg tasting tests, using free-range and battery cage eggs. With eggs that were

1–2 days old, 24 tasters recorded a preference for free-range eggs, against 13 for cage eggs. With 9–10-day-old eggs, 21 tasters preferred free-range against 16 for cage eggs. Differences in preference evened out as older eggs were tasted.[2]

Armed with as much information as possible about the product, the producer is in a good position to publicise it. The use of cheerful posters, leaflets and signs is not to be underestimated. Good packaging and labelling is also essential. There are regulations about this, which are examined later in this chapter, but as long as they are followed there is no reason why attractive labels should not be incorporated onto egg boxes. Many suppliers of egg packaging will arrange for the printing of your own logo if this is desired, or pre-printed labels which are stuck on can be utilised.

If sales are mainly local then local publicity is highly desirable. Sending a press release to the local newspapers and radio and television stations about your activities is one way of achieving this. If the enterprise includes a farm shop, some kind of attraction could be added. There are many possibilities. One could have a small play area, for instance, or a children's farm pet centre on the site. An attractive, welcoming place is far more likely to attract customers again if they are made to feel that their visit has been worthwhile. One free-range producer who ran a small survey among his customers to discover what they liked best was surprised to find that a frequent comment was 'the nice, clean toilets'!

If a press release is sent to the media it needs to have an interesting angle, such as the opening of an added attraction. It should be written in a clear, concise way. A suitable format is shown in Figure 10.2.

PACKAGING AND LABELLING

Eggs sold under the description free-range must be packaged in cartons, and all those who market eggs must be registered as an 'egg packing station' with the Egg Marketing Inspectorate at the nearest Ministry of Agriculture office. Small egg packs may use a 'sell-by date' which should be applied at the time of packing. The date ensures that the eggs will retain Grade A quality for a reasonable period of time after purchase by the consumer. The date may be shown in numbers or in a combination of numbers and letters, as long as it is clear to the consumer that it is a date by which the

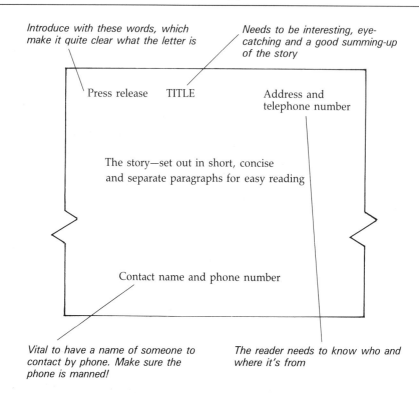

Introduce with these words, which make it quite clear what the letter is

Needs to be interesting, eye-catching and a good summing-up of the story

Press release   TITLE   Address and
                               telephone number

The story—set out in short, concise
and separate paragraphs for easy reading

Contact name and phone number

Vital to have a name of someone to contact by phone. Make sure the phone is manned!

The reader needs to know who and where it's from

Figure 10.2 How to produce a press release

eggs must be sold. For example, 'Sell by 20/06' or 'Sell by 20 June'. The term 'Best before . . .' is not permitted.

The Egg Marketing Inspector at the nearest MAFF/ADAS office will supply a copy of the European Community rules which govern egg sales. The address and telephone number is in the local telephone directory.

DISTRIBUTION

It depends entirely on the scale of operations whether or not a distributor is used. The small enterprise will tend to concentrate on local sales such as farm-gate sales, market stalls, local shops and hotels. The larger enterprise will need a distributor to collect the eggs on a regular basis. They will then be graded according to size and quality at the egg packing station, as described above, and distributed from there. A distributor, or one of the large retailing chains, will require the signing of a contract guaranteeing the quality of production and storage methods. Standards are usually stringent, including the regular inspection of sites, and the producer must be prepared to adopt high levels of management, efficiency and cleanliness. It is unlikely that a large distributor would collect small numbers of eggs at a time, so it is usually an option which is open to large enterprises only. Having said that, it is also true to say that where small producers have got together, a degree of co-operation has enabled them to appoint their own small-scale distributor.

One of the great problems facing the smaller producer is that he does not necessarily have the time to be both producer and distributor. It is an aspect which needs to be studied closely, for individuals will vary in their attitudes. What is certainly true is that if customers can be persuaded to call at the farm gate, the return is higher than with any other form of marketing.

Labeller

Figure 10.3 Examples of labelling

References

1 United Kingdom Egg Producers' Association, 1990.
2 *Farmers Weekly*, August 1987.

11 HEALTH

'Prevention is better than cure.'

The aim with all poultry is to avoid problems before they happen. This necessitates providing clean, well-managed housing and pasture, good food and clean water, and regular action to deter vermin and wild birds from the site. In addition there are some diseases which the birds should be protected against.

HOUSING

There are three 'Ds' to beware of in poultry housing – dirt, dust and droppings. All three harbour disease organisms and need to be dealt with on a regular basis. All obvious dirt should be cleaned away, and feeders and drinkers given particular attention. Nest box material should be checked regularly and replaced as necessary.

Every effort should be made to prevent dirt being brought into the house on birds' feet. The importance of providing a concrete, hardcore, slatted or porch area around the pop-holes was stressed earlier in the book.

Dust should also be kept to a minimum, and it is here that the intrinsic design of the house is important; one which provides a minimum of nooks and crannies that can harbour dust is advantageous. Ensuring that the ventilation is adequate is also a positive aid in keeping dust to a minimum.

Droppings obviously need to be removed frequently. If a droppings pit is used, it should be of sufficient depth to allow droppings to fall through the slats without allowing them to pile up in a heap which is accessible to the birds. Large perchery units would find the use of manure belts under the perches an advantage, but the small-to-medium house would probably rely on the use of droppings pits

144

or boards. The latter slide in and out under a slatted floor. Reference has already been made to the need to dispose of droppings away from the buildings and grazing areas, either by composting them for use as a horticultural manure, or removing them on a contract basis if the unit is a large one.

Buildings should have a periodic spring clean where all birds are removed, and the walls, ceiling and floor subjected to a thorough clean. Large houses will need power washers to do this, but small houses can be cleaned manually, paying particular attention to perches and nest boxes. A disinfectant or fumigant can be used to ensure that there are no bugs left lurking anywhere. The house should be left vacant for a couple of weeks before allowing birds to use it again.

PASTURE

Mention has been made several times of the need to ensure frequent changes of grazing in order to avoid a possible build-up of infection. Long grasses are hiding places for a range of infective organisms and keeping grass cut down is a positive way of minimising the risks of infection. Cutting also allows sunlight, which is a great cleanser, to have access to shorter grasses.

Pasture which is being rested will benefit from an application of lime which not only helps to 'sweeten' the soil, but contributes to breaking the life cycle of parasitic worms which may have infected the site.

THE BIRDS

Birds should be floor reared and have protection against Marek's disease, Newcastle disease and infectious bronchitis. It may also be wise to vaccinate them against endemic tremors if this is prevalent in particular localities. Veterinary or ADAS advice should be sought on this matter.

There are several agricultural colleges which offer training on a series of health-related topics. Of particular relevance are those which cover the procedures for administering vaccinations. Some of these are given in the drinking water, while others are by injection.

The latter should never be attempted without having had the technique taught by a professional.

Anyone who is breeding birds, no matter what the scale, should arrange for the breeding stock to be blood tested to ensure that they are healthy and free of blood and egg transmittable diseases. Again, the vet or ADAS will advise on this. Eggs which are to be incubated should be handled carefully and stored in cool conditions. Before incubation they should be dipped in an egg sanitant to ensure that they are as free of potentially infectious agents as possible.

Free-ranging birds are vulnerable to parasitic infection. Parasitic worms, for example, can be introduced by wild birds. If there is too great a burden of infestation, the chickens will need to be given an anthelmintic in the drinking water.

If the problem is one of external attack from lice, mites, ticks or fleas, an appropriate dusting powder is available from the vet. The birds, houses and dust bathing areas should all be treated. A particularly troublesome pest is the mite which causes scaly leg. This is a burrowing mite which produces encrustations which push up the scales of the legs. The best way of dealing with it is to bathe the legs in warm, soapy water and use an old toothbrush to dislodge the crusts. Do not try to remove them while the legs are dry, otherwise bleeding will result. Once the legs are clear and dry, spray them with benzyl benzoate, which is available from the vet. Ticks can be dealt with similarly, but on no account try to pull them away, otherwise the head will be left in, providing a source of infection. Spray it and leave it to drop off by itself.

All medications administered should be recorded, along with the date and dosage given. If eggs and table birds are being produced to organic standards, there is a requirement that medications should only be given where a problem arises. There are some medications which are not allowed as preventive vaccinations. Consult the appropriate standards which the distributor requires.

SALMONELLA TESTING

There is a requirement for anyone selling eggs (even from a single bird) or those with a breeding flock of 25 birds or more, to test the birds for salmonella. At present there are two ways of taking samples from laying birds – by means of swabs taken from the bird's vent, and from the faeces. The second is arguably less precise because of the possibility of salmonella organisms coming from an external source.

Some specialist laboratories provide sterile swabs, plastic bags, labels and instructions on how to take the swabs, free of charge. The procedure for the producer is then to take a swab, dip it into the peptone bottle provided and insert it into the vent of the bird. The swab is then dropped into the plastic bag provided and the procedure repeated with the other birds. The swabs are posted to the laboratory which will then test them and report back on their findings. If salmonella is detected, there is a legal requirement for the laboratory to inform the Ministry of Agriculture who will inspect the infected unit and, if deemed necessary, issue a compulsory slaughter order on the birds. This may not necessarily happen, if all that is required is to tighten up on general hygiene procedures, but the Ministry does have this power, a situation which is causing grave concern to the breeders of rare and traditional breeds in case some endangered species are completely wiped out.

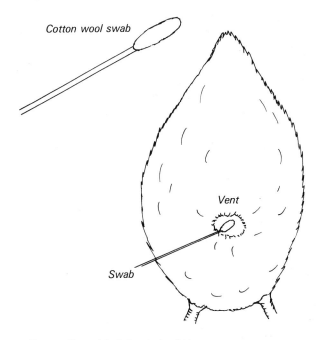

Source: Greendale Laboratories Ltd.

Figure 11.1 Method of taking a swab

147

It is interesting to note that some breeders have succeeded in being granted exemption on zoological grounds. In other words, if the breeds kept are reckoned to be of scientific value, the normal salmonella testing regulations do not apply. Exemptions should be applied for through the local Animal Health office.

The number of samples taken per laying flock will depend on its size. This is indicated in Table 11.1. Once birds have been introduced as point of lay pullets, the producer will need to test them once every 12 weeks. At the time of writing, the cost of doing this was £5.18 including VAT for 25 samples. The small poultry-keeper would therefore be paying a total of £20.72 including VAT per year. The large enterprise, testing a maximum number of 60 birds at £10.15 including VAT, 4 times a year, would pay a total of £40.60 including VAT.

Table 11.1   Salmonella testing requirements

Number of birds	Number of samples required
1–24	The total number kept
25–29	20
30–39	25
40–49	30
50–59	35
60–89	40
90–199	50
200–499	55
500 or more	60

One of the problems with some salmonella organisms is that they can be carried by apparently healthy birds, and passed directly into the egg, no matter how clean the environment may be. *Salmonella enteritidis* is an example of this. There is some evidence that increasing the acidic level of the feed inhibits salmonella in chickens. There are feed additives available which are made up of organic acids, notably formic and propionic acids, constituents which could hardly be frowned upon by the organic lobby. They are formulated into a granular material which is mixed with the normal feed, and in this form they inhibit salmonella and other pathogenic organisms.[2]

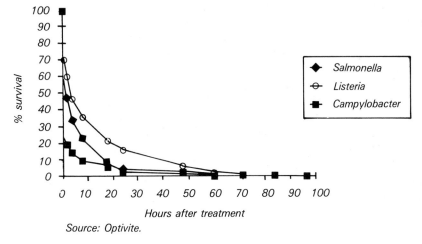

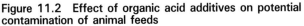

Source: Optivite.

Figure 11.2 Effect of organic acid additives on potential contamination of animal feeds

AN A–Z OF POULTRY DISEASES AND PROBLEMS

Aspergillosis (Fungal pneumonia, Pulmonary mycosis, Farmer's lung)

Caused by the fungus *Aspergillus fumigatus*, the condition is brought about by inhalation of the spores from contaminated litter or feed. Symptoms are excessive thirst, gasping and rapid breathing, with an overall depressed posture. Young birds are particularly at risk, and there is no effective treatment, although antibiotics have been shown to produce an improvement. Strict hygiene, avoidance of damp hay, straw, wood shavings and feed, together with good management of litter is needed. The condition can also affect man.

Avian encephalomyelitis (AE, Epidemic tremor)

This condition is caused by a virus which is transmitted primarily through the egg. The only effective control is by vaccination of breeders. The disease is seen mainly in chicks from 1–3 weeks old. Movements are restricted and trembling of the head and neck can be seen.

Breeding flocks which are affected will acquire an immunity, but no eggs should be incubated from them for several weeks until this is established.

Avian influenza

Avian influenza is caused by airborne viruses and affects the respiratory tract, as does the common cold in man. There may be a slight swelling of the head and neck, and a nasal discharge is usually seen. Mortality is normally low and there is no treatment. Antibiotics have been used where there are secondary infections. If the condition does not clear fairly rapidly, it should be investigated clinically for it can be confused with the more serious Newcastle disease, fowl pox and fowl plague.

Blackhead (Enterohepatitis, Histomoniasis)

More usually associated with turkeys, blackhead can also affect chickens. The causative agent is a protozoan parasite called *Histomonas meleagridis* which is transmitted via water, feed or droppings. Infected eggs of the parasitic caecal worm, *Heterakis gallinarum*, are also a source of infection. It is important not to allow ground to

Aspergillus spores. Low power electron micrograph of spore heads. (Reproduced courtesy of the School of Biological Sciences, University of Birmingham)

150

become overused, or to have chickens and turkeys sharing the same land.

Bumblefoot (Abscess of the foot)

Bumblefoot is the common name for the swelling which results from an infected cut or graze on the underside of the foot. The wound heals on the outside, leaving a hard core of pus on the inside. It is sometimes found where birds are provided with perches which are too high for them, or where the grazing area is on flinty ground. The affected bird will have a limp, and examination of the foot reveals the hard abscess. Applying slight pressure is sometimes enough to burst it, releasing the pus, but it may require lancing with a sterilised blade. Antiseptic liquid or cream should then be applied to the affected area.

Caecal worms (Heterakis)

Caecal worms are about 1.5 cm long and inhabit the caeca. They do not cause disease, although their eggs are capable of transmitting blackhead. They can be controlled by means of the vermifuge piperazine.

Cannibalism

Comparatively rare in well-managed free-range flocks, cannibalism is usually the result of stress, boredom, overintensification or lack of food. Further information is given in Table 6.1 on page 91.

Chronic respiratory disease (CRD, Airsacculitis)

Sneezing, coughing and wheezing are the signs which indicate this condition. It is initially caused by viral infection, followed by secondary bacterial invasion of the organism *Mycoplasma gallisepticum*. The best procedure is avoidance, by ensuring adequate ventilation in the house and a stress-free management system. Mild cases, where secondary infection is slight, will clear up fairly quickly, but severe cases may require antibiotic treatment.

As the tendency can be inherited via the egg, it is important to ensure that all breeding stock has been blood tested and found to be healthy. The symptoms are similar to those of Newcastle disease and infectious bronchitis, so if in doubt, consult a veterinary practitioner.

Cloactitis (Vent gleet, Poultry venereal disease)

The signs of this contagious venereal disease are swollen membranes in the cloaca, with a whitish discharge. The affected bird should be separated from the flock and the vent area painted with iodine. It may recover, otherwise it should be culled.

It is most commonly found in hens which are allowed to run indiscriminately with an infected male. The sexes should be kept quite separate unless controlled breeding is required, and the breeding stock has been checked for health.

Coccidiosis

Coccidiosis is caused by protozoan parasites which are transmitted via infected droppings. Affected birds are listless, have pale combs and pass blood in the droppings. Mortality can be as high as 50%. Coccidiocidal agents are available where there is an outbreak, but every effort should be made to change grazing areas frequently, and to deal with damp areas of litter.

Coryza (Infectious coryza, *Haemophilus gallinarum* infection, Roup)

Caused by the bacterium *Haemophilus gallinarum*, this is another condition which is best avoided by scrupulous attention to good ventilation in the house and general cleanliness in relation to equipment and environment.

Symptoms are similar to chronic respiratory disease, with inflammation of the eyes and nose, and sneezing and wheezing. Severe cases may require antibiotic treatment but mortality is generally low.

Egg drop syndrome (EDS)

The causative virus for this disease is transmitted through the egg. If carrier birds are subsequently incorporated into a flock, their droppings may provide a source of infection for the rest of the flock. Egg production is affected, with some deformed eggs being produced. The effect on egg production is shown in Figure 11.3.

There is no treatment, but a vaccination at the point of lay period provides protection.

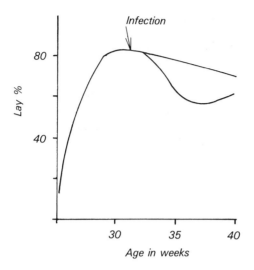
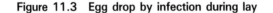

Source: Intervet Ltd.

Figure 11.3 Egg drop by infection during lay

Fowl cholera (*Pasteurellosis*)

Caused by the bacterium *Pasteurella multocida,* fowl cholera is recognised by swollen, bluish wattles in severely affected birds. It does not cause high mortality and can be treated with antibiotics in the drinking water. It is often carried by vermin so every effort should be made to deter them from the site.

Fowl pox (Avian pox, Avian diphtheria)

This is another virus infection transmitted by bird contact or via water and food. Mosquitoes are also known to be transmitting agents. There are characteristic lesions (pox marks) on the comb and there may also be laboured breathing. Mortality is usually low if the only symptoms are the lesions, but the more serious form which is accompanied by lung congestion causes around 40% mortality. If the disease is detected, it is advisable to vaccinate the whole flock without delay.

153

Fowl typhoid (Pullorum disease)

This is caused by the bacterium *Salmonella pullorum,* and affected birds produce white diarrhoea. Unless treated with antibiotics it can cause quite high mortality levels in the flock. It can also be transmitted via the egg, hence the importance of breeding only from blood tested, healthy birds.

Gumboro disease (Infectious bursal disease, IBD)

This is a highly infectious viral disease which is difficult to eradicate from a site. There is no treatment and breeding birds are normally vaccinated. It has been more apparent in the intensive broiler industry than in other areas of the poultry world, but it can affect all chickens.

Hairworms (Capillaria)

These internal worms cause damage to the lining of the intestine, resulting in anaemia and pale egg yolks. A suitable anthelmintic can be given in drinking water.

Infectious anaemia (Inclusion body hepatitis)

Transmitted via the egg and by droppings, this virus causes listlessness and some mortalities in the initial stages. If recovery takes place, the birds should not be allowed to breed. There is no treatment, although antibiotics can be used to control secondary infections.

Infectious bronchitis (IB)

This is a viral infection for which there is no treatment and chicks are normally vaccinated at an early stage. Where unvaccinated birds contact the disease any eggs laid will be deformed and, occasionally, devoid of shells.

Infectious laryngotracheitis (ILT)

Caused by a virus of the herpes group, ILT does not normally cause mortalities, but egg production in laying flocks may drop for about 3 weeks. Wheeziness and laboured breathing are signs of infection.

A vaccine is available to provide protection against it, but if scrupulous attention is paid to general hygiene, with adequate cleaning
of feeders and drinkers, there is little chance of birds encountering
it.

Lymphoid leucosis (LL, Big liver disease, Visceral leucosis)

This is a disease which is thought to be transmitted via the egg. It
is essential to breed only from breeders which are found to be clear
of it, following a blood test. Affected birds develop large tumours,
sometimes with enlarging of the bones and wings. There is no
cure.

Malabsorption syndrome (Pale bird syndrome, Stunting disease)

Little is known about this disease, but it is more frequently seen in
the broiler industry than in the free-range sector. Chicks develop
ricket-like symptoms, and have extremely pale heads and legs. There
is also diarrhoea with brown, foamy droppings. There is currently
no treatment. Hygienic conditions and fresh air with access to sun-
shine should prevent its appearance in outdoor birds.

Marek's disease (MD, Neurolymphomatosis)

This is another herpes virus disease which is transmitted through
the egg. Again, all breeders should be blood tested to ensure that
they are free of it before being allowed to breed. Paralysis and
tumours develop before death occurs. Day-old chicks are normally
vaccinated against it.

Newcastle disease (ND, Fowl pest)

This is a highly infectious disease caused by a paramyxovirus, and
poultry is normally vaccinated against it for there is no treatment.
Clinical signs are wheezing and choking and the onset of paralysis.
It is a notifiable disease, so the Ministry of Agriculture must be
informed.

Roundworms (Ascarids)

Found in the gut, these parasitic worms are around 5 cm in length, and although a healthy bird can tolerate a certain burden, if there are too many the result is anaemia and pale yolks. A preparation such as piperazine is effective in getting rid of them.

Tapeworms

Tapeworms grow to around 10 cm in the gut, but cause comparatively little harm unless the burden becomes too great. In this case, the bird needs to eat more than it would normally in order to maintain its metabolism. A suitable anthelmintic from the vet can be administered in severe cases.

References

1 Greendale Laboratories Ltd.
2 Optivite.

REFERENCE SECTION

REFERENCES

Specific references are detailed within the text or at the end of appropriate chapters.

OTHER PUBLICATIONS

Books

Practical Chicken Keeping, Katie Thear (Ward Lock).
British Poultry Standards, May & Hawksworth (Butterworth).
The Complete Handbook of Poultry-Keeping, Stuart Banks (Ward Lock).
Keeping Chickens, Walters & Parker (Pelham).
Poultry Health and Management, David Sainsbury (Granada).
Poultry and Poultry-Keeping, Alice Stern (Merehurst).
Bantams, Helga Fritzsche (Barron).
Incubation: A Guide to Hatching and Rearing, Katie Thear (Broad Leys Publishing).
Important Poultry Diseases (Intervet).
The Home Farm Sourcebook, D. & K. Thear (Broad Leys Publishing).
Salsbury Manual of Poultry Diseases (Salsbury Laboratories).
Modern Free Range, M. & V. Roberts (Domestic Fowl Trust).

Magazines

Fancy Fowl, Crondall Cottage, Highclere, Newbury, Berkshire. (Bi-monthly journal for the poultry fancier.)
Home Farm, Broad Leys Publishing Company, Buriton House, Station Road, Newport, Saffron Walden, Essex CB11 3PL. Tel: 0799 40922. (Small farmers' and smallholders' magazine published bi-monthly with regular articles on poultry.)
International Hatchery Practice, P.O. Box 4, Driffield, North Humberside YO25 8BJ. Tel: 026288 692. (Published eight times a year for those involved in commercial hatching and breeding.)

Poultry World, Carew House, Wallington, Surrey SM6 0DX. Tel: 081 647 4892. (Monthly journal for the commercial poultry industry.)

ORGANISATIONS

ADAS Poultry Group Headquarters, Great Westminster House, Horseferry Road, London SW1P 2AE.

ADAS Senior Poultry Advisor, Nobel House, 17 Smith Square, London SW1P 3HX. Tel: 071 238 3000.

British Chicken Information Service and British Egg Information Service, Bury House, 126–128 Cromwell Road, London SW7 4ET. Tel: 071 370 7411.

British Poultry Federation, 52–54 High Holborn, London WC1V 6SX. (For the commercial poultry producer. Also the headquarters of the British Chicken Association, British Egg Association and the British Poultry Breeders and Hatcheries Association.)

Compassion in World Farming, 20 Lavant Street, Petersfield, Hampshire GU32 3EW. Tel: 0730 64208.

Council for Small Industries in Rural Areas (COSIRA), 141 Castle Street, Salisbury, Wiltshire SP1 3TP.

The Farm Shop and Pick Your Own Association (FSPA), Agriculture House, Knightsbridge, London SW1X 7NJ. Tel: 071 235 5077.

Food from Britain, 301–344 Market Towers, New Covent Garden Market, London SW8 5NQ. Tel: 071 720 2144 (Egg quality scheme.)

Free Range Egg Association (FREGG), 37 Tanza Road, London NW3 2UA. (Group for support and promotion of free range egg production.)

Institute for Animal Health & Research, Houghton Laboratory, Houghton, Huntingdon PE17 2DA. Tel: 0480 64101. (Poultry health research.)

NAC Poultry Unit, NAC, Stoneleigh, Kenilworth, Warwickshire.

National Institute of Poultry Husbandry. Harper Adams Agricultural College, Edgmond, Newport, Shropshire TF10 8NB. Tel: 0952 820280.

National Farmers' Union (NFU) Agriculture House, Knightsbridge, London SW1X 7NJ. Tel: 071 235 5077.

The Poultry Club, Secretary: Mrs Liz Aubrey-Fletcher, Honce Farmhouse, Faircross, Stratfield Saye, Reading, Berks RG7 2BT.

Rare Poultry Society, Alexandra Cottage, 8 St Thomas's Road, Great Glenn, Leicestershire LE8 0EG.

Soil Association, 86–88 Colston Street, Bristol, Avon BS1 5BB. Tel: (0272) 290661. (Organic Symbol Scheme.)

United Kingdom Egg Producers' Association Ltd (UKEPRA), Brocklea, East Cundry, Bristol BS18 8NJ. Tel: 0272 643498. (UK egg producers' trade association.)

United Kingdom Register of Organic Food Standards (UKROFS), c/o Food from Britain, 301–344 Market Towers, New Covent Garden Market, London SW8 5NQ. Tel: 071 720 2144. (Organic symbol scheme.)

EDUCATIONAL ESTABLISHMENTS

Agricultural Training Board, National Poultry Office, York House, Clarendon Avenue, Leamington Spa CV32 5PP. Tel: 0926 421105.

Derbyshire College of Agriculture & Horticulture, Morley, Derby DE7 6DN. Tel: 0332 831 345.

Hadlow College of Agriculture & Horticulture, Tonbridge, Kent. Tel: 0732 850551.

Harper Adams Agricultural College, Edgmond, Newport, Shropshire TF10 8NB. Tel: 0952 820280.

Plumpton Agricultural College, Near Lewes, East Sussex BN7 3AE. Tel: 0273 890454.

Sparsholt College, Sparsholt, Winchester, Hampshire SO21 2NF. Tel: 0962 72441.

The West of Scotland Agricultural College, Auchincruive, Ayrshire KA6 5HW. Tel: 0292 520331.

West Sussex College of Agriculture & Horticulture, Brinsbury, Pulborough, West Sussex RH20 1DL. Tel: 07982 3832/3/4.

Worcester College of Agriculture, Hindlip, Worcester WR3 8SS. Tel: 0905 51310.

BREED CLUBS

Breed clubs are affiliated to the Poultry Club listed under Organisations.

Ancona Club, P. Smedley, Secretary, Beech Tree Farm, Flaxton, Yorks.

Appenzeller Club, D. Smillie, Secretary, Oak Drive Lodge, Bromsberrow HR8 1RY.

Araucana Club, Mrs P. Williams, Secretary, Bloomfield Cottage, Oakhill, Bath, Somerset.

Australorp Club, K. Sharpe, Secretary, 37 Bowgreave, Garstang, Preston, Lancashire.

Barnevelder Club, B. Clarke, Secretary, 113 Collingwood Road, Sutton, Surrey.

Black Wyandotte Club, E. Crossland, Secretary, 349 Manchester Road, Millhouse Green S30 6NQ.

Brahma Club, N. Senger, Secretary, 28 School Lane, Brereton, Sandbach CW11 9RN.

British Belgian Club, Miss V. Mayhew, Secretary, Behoes Lane, Woodcote, Reading, Berks.

Buff Orpington Club, D. Bruce, Secretary, 1 Letter Box Cottages, Bryants Bottom, High Wycombe, Bucks.

Croad Langshan Club, Mrs C. Hadley, Secretary, Deadmans Lane, Goring Heath RG8 7RX.

Derbyshire Redcap Club, S. Brassington, Secretary, Harthill Lodge, Harthill, Bakewell, Derbyshire.

Dutch Bantam Club, Mrs S. Parson, Secretary, Goring Road, Woodcote, Reading, Berkshire.

Faverolles Society, Mrs S. Bruton, Secretary, Codsall Wood, Staffordshire WV8 1QR.

Frizzle Society, Mrs E. Amos, Secretary, 53 Cambridge Street, Loughborough LE11 1NL.

Hamburgh Club, H. Critchlow, Secretary, Little Longsdon Farm, Longsdon, Stoke-on-Trent, Staffordshire.

Indian Game Club, J. Cook, Secretary, 15 Campden Road, Gravenhurst, Bedfordshire.

Japanese Club, M. Fowler, Secretary, Cockhill Cottage, Cockhill, Castle Cary, Somerset.

Laced Wyandotte Club, A. Askew, Secretary, 14 Lanesfield Park, Evesham, Worcestershire.

Leghorn Club, B. Turner, Secretary, Barnham Lane, Walburton BN18 0AY.

Maran Club, Mrs A. Rolls, Secretary, East Dunley, Bovey Tracey, Newton Abbot TQ13 9PW.

Minorca Club, P. Gray, Secretary, 1 Kielder Salmon Hatchery, Hexham NE48 1HX.

Modern Game Bantam Club, A. Higgs, Secretary, Manor Farm, Wingfield, Leighton Buzzard LV7 9QH.

Orpington Club, Mrs H. J. Bridson, Secretary, 4 Arthog Road, Didsbury, Manchester.

Partridge Wyandotte Club, G. A. Parker, Secretary, Saunton, The Runnell, Neston, Wirral, Cheshire.

Poland & Poland Bantam Club, R. Berry, Secretary, 7 Archgrove, Long Ashton, Bristol.

Rhode Island Red Club, M. Froggatt, Secretary, Eastern Lodge, Cornwood Road, Plympton, Plymouth.

Rosecomb Bantam Club, R. E. Sharpe, Secretary, Broadfield, Pilling, Preston, Lancashire.

Scots Grey Club, J. Robertson, Secretary, 21 Kerswell Avenue, Kaimend, Carnwath ML11 8LE.

Sebright Club, K. Sharpe, Secretary, 37 Bowgreave, Garstang, Preston, Lancashire.

Silkie Club, Mrs D. Ryer, Secretary, Lower Vicarwood Farm, Mackworth, Derbyshire.

Sussex Club, M. Raisey, Secretary, Ashford Hill Road, Headley, Newbury, Berkshire.

Welsummer Club, Mrs Bullen, Secretary, Sandyford Lane, Old Leeke, Boston PE22 9RB.

White Wyandotte Club, A. Procter, Secretary, 2 Manor Avenue, Ribchester, Preston, Lancashire.

Wyandotte Bantam Club, Miss J. Newton, Secretary, 93 Steam Mill Lane, Ripley, Derbyshire DE5 3JR.

SUPPLIERS

Housing

ADF, Unit 5, Leasowe Green, Lightmoor Village, Telford TF4 3QX. Tel: 0952 592309. (Range of arks.)

Simon Bowler (Sectional Buildings), Elmhurst, Egginton, Derby DE6 6HQ. Tel: 028 373 3947.

CEF Agricultural (Holdings) UK, 2 Pinfold Lane, Scarisbrick, Ormskirk, Lancashire L40 8HR. Tel: 0704 840 980. (Also equipment.)

Challow Products Agriculture Ltd, Unit 4, Old Sawmills Road, Faringdon, Oxfordshire SN7 7DS. Tel: 0367 20091. (Large and small houses for commercial applications.)

Chapelfield Poultry, Coventry Road, Dunchurch, Rugby, Warwickshire. Tel: 0788 810111. (Also incubators, equipment and supplies.)

J. Davies, Woodland Farm, Hinstock, Market Drayton, Shropshire TF9 2TA. (Plans for 12-hen fold unit.)

Domestic Fowl Trust, Honeybourne, Evesham, Worcestershire WR11 5QJ. Tel: 0386 833083. (Also general supplies.)

Forsham Cottage Arks, Goreside Farm, Great Chart, Ashford, Kent TN26 1JU. Tel: 023 382 229.

Barry Fowler Timber Products, Morrishes Farm, West Buckland, Wellington, Somerset. Tel: 0823 47 5350.

Gardencraft, Ian and Chris Shutes, Tremadog, Porthmadog, Gwynedd LL49 9RD. Tel: 0766 513036.

Hardwick Farms, Horsecroft, Hardwick Lane, Bury St Edmunds, Suffolk IP29 5NY. Tel: 0284 4611.

A. E. Jennings, Iron Cross, Salford, Priors, Evesham WR11 5SH. Tel: 0386 870321. (Also incubators, equipment and electric netting.)

G. Longhurst, Whitegates, Foulden, Norfolk IP26 5AQ. (DIY plans for all-weather bantam unit.)

D. Mears, The Old Mill, Buckingham Road, Brackley, Northants NN13 5AL. Tel: 0280 703804.

Meirionnydd Woodworkers' Cooperative Ltd, 1 Glan y Don, High Street, Barmouth, Gwynedd LL42 1DW. Tel: 0341 423345.

Newbridge Farm, Aylton, Ledbury, Herefordshire. Tel: 053 183 276/676. (Also general equipment.)

Park Lines (Buildings) Ltd, Park House, 501 Green Lanes, Palmers Green, London N13 4BS. Tel: 081 886 0011.

Penstrowed Poultry, College Farm, P.O. Box 2, Newtown, Powys SY16 1ZZ. Tel: 0686 28582.

Pentangle Enterprises, Wing of Southcott Farm, Cotleigh, Honiton, Devon. Tel: 040 483 420.

John Price & Sons, New Street, Ledbury, Herefordshire. Tel: 053 183 676. (Large houses, equipment and fencing.)

SPR Poultry Centre, Barnham, Bognor Regis, Sussex. Tel: 0243 554006/ 542815.

Summit Supplies and Services Ltd, Briery House, 134 Blackgate Lane, Tarleton, Preston, Lancashire. Tel: 0772 813111. (Also nest boxes.)

Ten Hen Ltd, Robin Clover, The Gables, Framlingham Pigot, Norwich, Norfolk NR14 7QJ. Tel: 0586 2453. (Small flatpack house.)

Wallace Fox, 12 Ploughley Close, Ardley, Bicester, Oxfordshire. Tel: 0869 345437.

Walner Wildfowl, Cwmkesty Farmhouse, Newchurch, Kington, Herefordshire. Tel: 0544 22603.

Wigfield & Pluck Ltd, Buckle Mill, Honeybourne, Nr Evesham, Worcs. (Large-scale housing.)

Woodland Ways, Elm House, Main Road, Saltfleetby, Louth, Lincolnshire. Tel: 050 783 230.

Polythene Houses

John R. Allright Ltd, New House, Longdon, Tewkesbury, Glos.
GL20 6AR. Tel: 068 481 445.

FRI, College House, Penstrowed, Newtown, Powys SY16 1ZZ. Tel:
0686 28582.

Polybuild, Unit 9C Tewkesbury Industrial Centre, Delta Drive,
Tewkesbury, Gloucestershire GL20 8HB. Tel: 0684 297109.

Polygrow, Unit 1, Bacton Mill Stores, Spa Common, North Walsham,
Norfolk. Tel: 0692 403665.

Roofing and Cladding

Brohome Ltd, Bedwas House Industrial Estate, Bedwas, Gwent NP1
8DW. Tel: 0222 860166.

Onduline, Ofic (GB) Ltd, Eardley House, 4 Uxbridge Street, Farm Place,
Kensington, London W8 7SY. Tel: 071 727 0533.

Insulation Services

Baxenden Chemicals Ltd, Paragon Works, Baxenden, Near Accrington,
Lancashire BB5 2SL. Tel: 0254 872278. (Sprayed urethane.)

British Insulations, Mill View, Shouldham Thorpe, King's Lynn, Norfolk
PE33 0EB. Tel: 03664 401. (Sprayed urethane.)

Electric Fencing

Bramley & Wellesley, Unit C, Chancel Close Trading Estate, Eastern
Avenue, Gloucester GL4 7SN. Tel: 0452 300 450.

Drivall Ltd, Narrow Lane, Halesowen, West Midlands B62 9PA. Tel:
021 421 7007.

Renco, Unit K1A, Bath Road Trading Estate, Stroud, Gloucestershire
GL5 3QF. Tel: 0453 72154.

Feeding Systems

Autonest Ltd, Brampton Wood Lane, Desborough, Northants
NN14 2SR. Tel: 0536 760332.

Biodesign Products Co., 15 Sandyhurst Lane, Ashford, Kent. Tel:
0233 26677. (Outside auto-feeders.)

163

Broiler Equipment Co. Ltd, Moorside Road, Winnal, Winchester, Hampshire SO23 7SB. Tel: 0962 61701. (Also drinking systems.)

EB Equipment Ltd, Redbrook, Barnsley, Yorkshire, S75 1HR. Tel: 0226 206896.

George H. Elt Ltd, Eltex Works, Worcester. Tel: 0905 422377. (Also drinkers.)

Drinking Systems

Lubing Equipment (UK) Ltd, Wellington Close, Parkgate Industrial Estate, Knutsford, Cheshire WA16 8DX. Tel: 0565 50207.

Rainbow Valve Co. Ltd, Upthorpe Road, Stanton, Bury St Edmunds, Suffolk IP31 2AU. Tel: 0359 50238.

Feeds

R. & E. Bamford, The Corn Mill, Bretherton, Preston, Lancashire PR5 7BD. Tel: 0772 600671.

J. Bibby Agriculture Ltd, Oxford Road, Adderbury, Banbury, Oxfordshire OX17 3HL. Tel: 0295 810281.

BOCM Silcock Ltd, Basing View, Basingstoke, Hampshire RG21 1SB. Tel: 0256 843210.

Dalgety Agriculture Ltd, 180 Aztec West, Almondsbury, Bristol BS12 4TH. Tel: 0454 201511.

Poultry Processing

Bingham Appliances (I.B.) Ltd, 65 Beaufoy Road, Tottenham, London N17 8AX. Tel: 081 801 1156. (Plucking equipment, Sinupul for removing leg tendons and humane poultry killer.)

Cope & Cope Ltd, Vastern Road, Reading, RG1 8BX. Tel: 0734 54491. (Killing and plucking equipment.)

Minting Farm Supplies Ltd, Minting, Horncastle, Lincolnshire LN9 5RX. Tel: 065 887 220. (Coloder plucking powder.)

Warren Agricultural Machinery, Warren Farm Cottage, Headley Lane, Mickleham, Dorking, Surrey RH5 6DG. Tel: 0372 377279. (Humane killers.)

Packaging

Boxmore Ltd, Boxmore Works, Inn Road, Dollingstown, Lurgan BT66 7JW. Tel: 0762 327711. (Egg trays and packaging.)

Clearprint, St Thomas Place, Preston PR1 1JU. Tel: 0772 58185. (Labels for pre-packed eggs.)

Danro Ltd, Unit 68, Jaydon Industrial Estate, Station Road, Earl Shilton, Leicester LE9 7GA. Tel: 0455 847061/2. (Labels and labellers for egg producers.)

Omni-Pac UK Ltd, South Denes, Great Yarmouth, Norfolk NR30 3QH. (Egg cartons and trays.)

Scandinavian Packing Co. (UK) Ltd, Exchange House, Exchange Square, Beccles, Suffolk NR34 9HH. Tel: 0502 717101. (Egg trays and cartons.)

Nest Boxes

Anglian Livestock Appliances Ltd, Lower Happisburgh, Norwich, Norfolk NR12 0AJ. Tel: 0692 650497.

Clifford Kent Ltd, Horizon Farm, Tremar, Liskeard, Cornwall. Tel: 0579 45938. (Poultry equipment and nest boxes.)

Equipment for Livestock Management Ltd, Great Doddington, Northamptonshire NN9 7TA. Tel: 0933 223278.

Harlow Bros Ltd, Long Whatton, Loughborough, Leicestershire LE12 5DE. Tel: 0509 842561.

George Mixer & Co., Catfield, Great Yarmouth, Norfolk NR29 5BA. Tel: 0692 80355.

R. J. Patchett Ltd., Ryefield Works, Clayton Heights, Queensbury, Bradford, West Yorkshire.

Poultry and Livestock Equipment, 99 Colchester Road, West Bergholt, Colchester, Essex CO6 3JX. Tel: 0206 241867. (Rollaway nest boxes and general equipment.)

Sprior (GB) Ltd, 4A Atcham Industrial Estate, Shrewsbury, Shropshire SY4 4UG. Tel: 0743 75891.

Spirofeed Ltd, Autertown Industrial Estate, Mallow, Co. Cork, Ireland. Tel: 022 21803 21727. (Free-standing nest boxes.)

Sterling Farm Equipment, Cod Beck Estate, Dalton, Thirsk, North Yorkshire Y07 3HR. Tel: 0845 577811.

Veterinary Products and Services

Glaxo Animal Health Ltd, Breakspear Road South, Harefield, Uxbridge, Middlesex UB9 6LS. Tel: 089 56 30266.

Greendale Laboratories Ltd, Lansbury Estate, Knaphill, Woking, Surrey GU21 1EW. Tel: 0483 797707. (Salmonella and other testing.)

Intervet UK Ltd, Science Park, Milton Park, Cambridge CB4 4FP. Tel: 0223 962751.

Kaycee Veterinary Products Ltd, 8 Park Road, Haywards Heath, West Sussex RH16 4HZ. Tel: 0444 452300.

Micro-Biologicals Ltd, Fordingbridge, Hampshire SP6 1AE. Tel: 0425 52215.

Optivite Ltd, Main Street, Laneham, Retford, Nottinghamshire DN22 0AN. Tel: 077 785 741. (Traditional fancy fowl pellets treated with Salkil to protect birds from salmonella infection.)

Peter Hand Animal Health, 15–19 Church Road, Stanmore, Middlesex, HA7 4AR. Tel: 081 954 7422. (Also consultancy service.)

Salsbury Chemicals Ltd, Solvay House, Flanders Road, Hedge End, Southampton SO3 4QH. Tel: 04892 81711.

Incubators

A. B. Incubators Ltd, 40 Old Market Street, Mendlesham, Suffolk IP14 5SA.

D. E. Bonnett, Longcroft, Dowset Lane, Ramsden Heath, Billericay, Essex. Tel: 0268 710197. (Also general equipment.)

Brinsea Products Ltd, Station Road, Sandford, Avon BS19 5RA. Tel: 0934 823039.

Bristol Incubators, Game Farm, Latteridge, Iron Acton, Bristol BS17 1TY.

Curfew Incubators, Southminster Road, Althorne, Essex CM3 6EN. Tel: 0621 741923.

Eco Electrics, Unit 1, Bosleake, Redruth, Cornwall TR15 3YG. Tel: 0209 612264. (DIY incubator/brooder kits. Electronic thermostats for incubators.)

George H. Elt Ltd, Eltex Works, Worcester WR2 5DN. Tel: 0905 422377. (Also brooders and general equipment.)

Mardle Products, Beechwood, Postbridge, Yelverton, Devon. Tel: 0822 88252.

Meadows Poultry Supplies, 12 Peterborough Road, Castor, Peterborough PE5 7AX. Tel: 0733 380288. (Also general equipment.)

Pintafen Ltd, 93 Hospital Road, Bury St Edmunds, Suffolk IP33 3LH. Tel: 0284 752828. (Also general equipment.)

Smallholding and Farm Supply Co., Toledo Works, Holliscroft, Sheffield S1 4BG. Tel: 0742 700651. (Also general equipment.)

Smallholding Supplies, Pikes Farmhouse, East Pennard, Shepton Mallet, Somerset BA4 6RR. Tel: 074 986 688. (Also general equipment.)

Brooders

Maywick (Hanningfield) Ltd, Rettendon Common, Chelmsford Essex.
EMC Gas Appliances, Anson Road, Martlesham Heath Industrial Estate,
  Ipswich, Suffolk IP5 7RG. Tel: 0473 625151.

General Equipment and Supplies

Cyril Bason (Stokesay) Ltd, Craven Arms, Shropshire SY7 9NE. Tel:
  05882 3204/5 or 3242. (Also vaccines and egg packaging.)
Baty International Ltd, Victoria Road, Burgess Hill, West Sussex
  RH15 9LB. Tel: 04446 5621. (Haugh unit gauges.)
George H. Elt Ltd, Eltex Works, Worcester WR2 5DN. Tel: 0905 422377.
Robert Frazer & Sons Ltd, Hebburn, Tyne & Wear NE31 1BD. Tel:
  0632 84616. (Tracer self-regulating heating tape for water pipes.)
The General Chip Company, Denver Site, Ferry Lane, Rainham, Essex.
  Tel: 04027 23035. (Wood shavings. Nationwide distribution points.)
R. & G. Gwilt Engineering, Waterworks Lane, Leominster,
  Herefordshire. Tel: 0568 3663.
Lincolnshire Smallholders' Supplies Ltd, Thorpe Fendykes, Wainfleet,
  Lincolnshire PE24 4QH. Tel: 075 486 255.
Smallholding Supplies, Pikes Farmhouse, East Pennard, Shepton Mallett,
  Somerset BA4 6RR. Tel: 074 986 688.
Snowflake Woodshaving Co. Ltd, Marsh Lane, Boston, Lincolnshire
  PE21 7ST. Tel: 0205 57015. (Available from six nationwide depots.)
Sorex Ltd, St Michael's Industrial Estate, Widnes, Cheshire WE8 8TJ. Tel:
  051 420 7151. (Bait for eradicating mice and rats.)
West Dorset Bird Farms, Corscombe, Dorchester DT2 0PA. ('Electric Egg'
  to deter poultry from pecking eggs.)
Woodside Farm and Wildfowl Park, Mancroft Road, Slip End, Luton,
  Bedfordshire. Tel: 0582 841044.

Seeds for poultry leys

MAS, W. Evans, 5 Brevel Terrace, Charlton Kings, Cheltenham
  GL53 8JZ. Tel: 0242 34255.

Buyers of non-intensive table poultry

Real Meat Company, East Hill Farm, Heytesbury, Warminster, Wilts.
  BA12 0HR. Tel: 0985 40436.

167

STOCK

Pure Breeds

David Applegarth, Uplands, Walkington, Beverley, Humberside
HU17 8SP. Tel: 0482 860862. (Autosexing breeds, Legbars.)

D. E. Bonnett, Longcroft, Dowset Lane, Ramsden Heath, Billericay,
Essex. Tel: 0268 710197. (Range of utility breeds.)

Chapelfield Poultry, Coventry Road, Dunchurch, Rugby, Warwickshire.
Tel: 0788 810111. (Marans, Welsummers, RIRs, Light Sussex and
Minorcas.)

Daviot Stud, Miss A. L. Clark, Tudor House, Nuthurst, Hockley Heath,
Warwickshire. Tel: Lapworth 3254. (Pedigree Large fowl and Light
Sussex bantams.)

The Domestic Fowl Trust, Honeybourne, Evesham, Worcestershire
WR11 5QJ. Tel: 0386 833083. (Range of utility breeds, rare breeds and
bantams.)

Roland and Sally Doughty, Aberfoyle, School Lane, Broadholm, Saxilby,
Lincoln LN1 2LZ. Tel: 0522 702934. (Welsummers, Marans, RIRs, Light
Sussex, Wyandottes and Orpingtons.)

F. J. Kitchen, Heathcote Grange, Heathcote, Hartington, Buxton,
Derbyshire. Tel: 029884 210. (RIRs, Leghorns, Wyandottes and
Derbyshire Redcaps. Also bantams.)

Knighton Free Range Poultry, Mr & Mrs E. F. J. Akehurst, Westowe
Manor Farm, Lydeard St Lawrence, Taunton, Somerset. Tel: 09847 206.
(Silkies, Sultans, Belgian fowl.)

Novoli Poultry Breeding Farm, The Street, Purleigh, Essex. Tel:
0621 828262. (Marans, Welsummers, RIRs, Light Sussex, Buff Rocks,
White Leghorns, Black Minorcas, Australorps, Anconas – chicks and
growers.)

Pennine Poultry, Old School Farm, Uppertown, Ashover, Chesterfield
S45 0JF. Tel: 0246 590813. (Marans, Barnevelders, RIRs, Light Sussex.)

Richard A. Rowley, 98 Sutton Road, Kirkby-in-Ashfield,
Nottinghamshire. Tel: 0623 755423. (Orpingtons, Australorps and
Wyandottes.)

Jim Self, Sycamore Cottage, Fishtoft Drove, Foston, Lincolnshire
PE22 7ES. Tel: 0205 750261. (Orpingtons, Wyandottes. Also range of
bantams.)

SPR Poultry Centre, Barnham Station Car Park, Barnham, Bognor Regis,
Sussex. Tel: 0243 554006/542815. (Domestic pure breeds including
Marans, Welsummers, Light Sussex and Rhode Island Reds.)

The Wernlas Collection, Green Lane, Onibury, Near Ludlow,
Shropshire SY7 9BL. Tel: 058 477 318. (Extensive collection of breeds.)

Hybrid Layers and Broilers

Tom Barron Hatcheries Ltd, Catforth, Preston, Lancashire PR4 0HQ. Tel:
0772 690111. (ISA brown layers.)

Cyril Bason (Stokesay) Ltd, Craven Arms, Shropshire SY7 9NG. Tel:
0588 673204/5 or 673242.

The Cobb Breeding Company Ltd, East Hanningfield, Chelmsford, Essex
CM3 8BY. Tel: 0245 400109. (Cobb broilers.)

International Poultry Services Ltd, Green Road, Eye, Peterborough,
Cambridgeshire PE6 7YP. Tel: 0733 223333.

Johnstone of Mountnorris, 11 Porthill Road, Mountnorris, Armagh
BT60 2TY. Tel: 086 157 281 or 0387 52667.

Joice & Hill Ltd, South Raynham, Fakenham, Norfolk. Tel: 032874 261
and 0284 810243. (Hisex Brown layers.)

Maple Leaf Chicks Ltd, 55/57 Garstang Road, Preston, Lancashire
PR1 1LB. Tel: 0772 24534.

Maurice Millard (Chicks) Ltd, Peipards Farm, Freshford, Bath BA3 6DN.
Tel: Limpley Stoke 2215.

Ross Poultry Ltd, The Broadway, Woodhall Spa, Lincolnshire LN10 6PS.
Tel: 0526 52471.

Sappa Chicks, The Hatchery, Fornham-All-Saints, Bury St. Edmunds,
Suffolk IP28 6JJ. Tel: 0284 752233.

Shaver Poultry Breeding Farms (GB) Ltd, Elsing Lane, Bawdeswell,
Dereham, Norfolk NR20 4QH. Tel: 036 288 254.

South Western Chicks (Warren) Ltd, Broadway, Ilminster, Somerset. Tel:
04605 3441.

R. Thompson (Chicks) Ltd, Lannhall, Tynron, Thornhill, Dumfrieshire,
Scotland. Tel: 08482 329.

Victoria Hatchery, Victoria Road, Diss, Norfolk IP22 3HE. Tel: 0379 2356.

Whitakers Hatcheries Ltd, Camden Quay, Cork, Eire. Tel: Cork 503366.

Poultry Collections on View

The Domestic Fowl Trust, Honeybourne, Evesham, Worcestershire
WR11 5QJ. Tel: 0386 833083.

The Wernlas Collection, Green Lane, Onibury, Near Ludlow, Shropshire
SY7 9BL. Tel: 058 477 318.

Folly Farm, Duckpool Valley, Near Bourton-on-the-Water, Cheltenham,
Glos. Tel: 0451 20285.

INDEX

Figures in italics indicate tables, figures or illustrations

171

Farming Press Books

Below is a sample of the wide range of agricultural and veterinary books published by Farming Press. For more information or for a free illustrated book list please contact:

Farming Press Books, 4 Friars Courtyard
30–32 Princes Street, Ipswich IP1 1RJ, United Kingdom
Telephone (0473) 241122

Outdoor Pig Production *by Keith Thornton*
How to plan, set up and run a unit

Goat Farming *by Alan Mowlem*
Covers all aspects for those considering goats an as alternative enterprise

Organic Farming *by Nicholas Lampkin*
Principles and practice for livestock and crops

Pearls in the Landscape *by Chris Probert*
The creation, construction, restoration and maintenance
of farm and garden ponds for wildlife and countryside amenity

Farm Woodland Management *by Blyth, Evans, Mutch and Sidwell*
Covers the full range of woodland size
from hedgerow to plantation with the emphasis on economic benefits
allied to conservation

Farming Press also publish four monthly magazines: *Livestock Farming, Arable Farming, Dairy Farmer* and *Pig Farming.* For a specimen copy of any of these, please contact Farming Press at the address above.